John Seller

A New Systeme of Geography

Designed in a most plain and easie method, for the better understanding of that science: accommodated with new maps, of all the empires, kingdoms, principalities, dukedoms, provinces and countries in the whole world

John Seller

A New Systeme of Geography

Designed in a most plain and easie method, for the better understanding of that science: accommodated with new maps, of all the empires, kingdoms, principalities, dukedoms, provinces and countries in the whole world

ISBN/EAN: 9783337244057

Printed in Europe, USA, Canada, Australia, Japan

Cover: Foto ©berggeist007 / pixelio.de

More available books at **www.hansebooks.com**

A NEW SYSTEME:
OF
Geography,

Designed in a Moſt Plain and Eaſy Method, for the better Underſtanding of that Science.

Accommodated With

NEW MAPPS,

OF ALL THE

Countreys, Regions, Empires, Monarchies, Kingdoms, Principalities, Dukedoms, Marqueſates, Dominions, Eſtates, Republiques, Soveraignties, Governments Seignories, Provinces, and Countreys in the whole World.

WITH

Geographical Tables,

Explaining the Diviſions in Each Mapp.

By *John Seller*, Hydrographer to the King.

And are Sold at his Shop on the Weſt-ſide of the *Royal Exchange*.

A New SYSTEME
OF
Geography.

The Elements and Principles of Geography.

CHAP. I.
Of the Nature and Division of Geography.

EOGRAPHY is a Science shewing the Divisions and Distinctions of the Earthly Globe, as it is a Spherical Body, Composed of Earth and Water, for that both these do together, make one Globe.

2. And hence the Parts of Geography are two, the one concerns the Earthy, and the other the Watry part.

3. The Earthy part of the Globe may be divided into Continents and Islands.

4. A Continent is a great quantity of Land, not Separated by any Sea, from the rest of the World; as the whole Continent of *Europe*, *Asia* and *Africa*.

5. An Island is a part of Land Environed with some Sea or other, as the Islands of *Great-Britain*, and *Ireland*, with the Ocean : The Island of *Sicilia* with the *Mediterranear*.

6. Both those are Subdivided into *Peninsula*, *Isthmus*, *Promontorium*.

7. *Peninsula* is a Tract of Land which being almost Encompassed round by Water, is joined with some little part or neck of Land.

8. *Isthmus* is that narrow neck of Land which joineth the *Peninsula* to the Continent, thus is *Africa* joined to *Asia*, only by that small neck of Land that is Contained between the *Mediterranean*, and the *Arabian* Gulf; called the *Red Sea*.

9. *Promontorium* is a high Mountain which shooteth it self into the Sea, the utmost end thereof is called a Cape, as Cape *Bona Esperance* in *Africa*.

10. The Watery part of the Globe is Destinguished by Divers Names, as *Oceanus*, *Maro*, *Fretum*, *Sinus*, *Lacus* and *Fluvius*.

1. *Oceanus*, or Ocean, is that Great general Collection of Waters that Encompasseth the Earth on every side.

2. *Mare*, the Sea, is a part of the Main Ocean, to which we cannot come but through some *Fretum* or Strait, as *Mare Meditterraneum*, and sometimes takes its Name from the Adjacent Shore; as *Mare Adriaticum*, from the City of *Adria*, or from a first Discoverer, as *Mare Magelanicum*.

3. *Fretum*

3. *Fretum* or Strait, Is a part of the Ocean, straitned between some Narrow bounds, and opening the Way to some Sea, as the Straits of *Gibralter*, the Straits of *Magelan*, &c.

4. *Sinus* Is a Gulf or Bay or any Indraught of Water as the Gulf of *Venice*, the Gulf of *Mexico*, the Bay of *Biscaia*, and the Bay of *Bengale* in the *East-Indies*.

5. *Lacus* or a Lake, is a Body or Collection of Waters, which hath no visible Intercourse with the Sea, or Influx into it, as the Lake of *Geneva*, and the Lake of *Asphaltites*, or Dead Sea, in the Land of *Canaan*.

6. *Fluvius* or River, is a Water-course Issuing from some Spring or Lake. —— And Emptyeth it self into some part of the Sea, a great River, as the Rhine, the Thames, &c.

CHAP. II.
Of the Circles of the Sphere.

THere are ten Circles of the Sphere, six great: and four lesser.

The Six great Circles, are the Meridian, the Horizon, the Equinoctial, the Ecliptick, the two Colures, all which divide the Sphere into two Equal parts.

The four lesser Circles are the two Tropiques, and the two Polar Circles, each of which Divides the Sphere into two unequal parts.

The *Meridian* is a great Circle which passeth through both the Poles of the World, and through the *Zenith* and *Nadir* Points, and sheweth the Latitudes of all places on the Earth.

The *Horizon* is also a great Circle which divideth the visible part of the Heavens, which we see, from those we see not.

The *Equinoctial* is a great Circle 90 Degrees from either Pole, in which Circle are reckoned the Longitude of all places on the Earth, from any certain Meridian Assigned, which Primary Meridian (from whence the Longitude of places in all the Mapps in this Treatise, doth Commence,) is that which passeth through the Island of *Pico Teneriffa*, and from thence Reckoned Eastward round the World.

The *Ecliptique* is a great Circle divided by the Equinoctial into two Equal Parts, one declining towards the North, and the other towards the South, the greatest Obliquity being 23 Degrees, 30 Minutes.

The Colures are two great Circles of the Sphere Intersecting each other at Right Angles in the Poles of the World. One is called the *Solstitial*, the other the Equinoctial Colure: The Solstitial Colure is that which passeth through the Poles of the World, and cutteth the Ecliptique in the Points of *Cancer*, and *Capricorne*.

The Equinoctial Colure passeth through the Poles of the World, and cutteth the Ecliptique, and the Equator in the Points of *Aries* and *Libra*, by which Points the four Seasons of the Year are Distinguished.

The

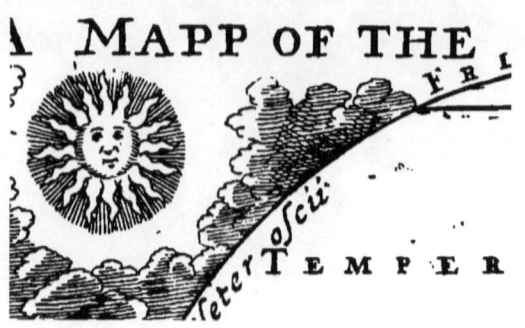

of Geography.

The Four Lesser Circles of the Sphere.

The four lesser Circles of the Sphere are the two Tropiques and the two Polar Circles, which Circles Divides the Earth into five broad Spaces, called Zones, which are distinguished in the following Chapter.

CHAP. III.

Of the Division of the Earth by Zones.

A Zone is a broad Space of the Earth limited by the Tropiques, and the Polar Circles, and are five in Number, one Torrid (or burning Zone) two Temperate, and two Frozen Zones.

1. The Torrid or Burning Zone is, that Space of Earth and Sea Contained between the two Tropiques, and is 47 Degrees in Breadth.

2. The two Temperate Zones are that Space Contained between each Tropique, and the Polar Circles, one called the North, and the other the South Temperate Zone: And are each of them 43 Degrees in Breadth.

3. The two Frozen Zones are those Spaces Contained between each Polar Circle, and the Poles of the world.

4. The Inhabitants of those Zones in Respect of the Diversity of their Noon Shadowes, are Divided into three Kindes, *Amphiscii, Heteroscii,* and *Periscii.*

5. Those that inhabit in the Torrid Zone are called *Amphiscii*, because their Noon-shadows are Diversly cast, sometimes towards the North, and sometimes towards the South, according to the Position of the Sun, when it is on the North or South side of their Zenith, or Vertical Point; and their Shadows are cast accordingly, *viz.* If the Sun be to the Northward, then their Shadow is cast to the Southward, and so on the Contrary.

6. Those Inhabitants that live in the Temperate Zones are called *Heteroscii*, because their Noon Shadows are cast but one way, and that either North or South; Those who live in the North Temperate Zone, their Noon Shadows are alwaies cast towards the North, and those of the South Temperate Zone, their Noon Shadows are alwayes cast towards the South. As may be seen in the Anexed Mapp of the Zones.

7. Those who Inhabit in the Frozen Zones, are called *Periscii*, because their longest day is at least 24 hours in length, and therefore the Sun being Carryed Circularly about them, their Shadows are also Carryed about them, in the same manner.

CHAP. IV.

Of the Division of the Earth by Climates.

A Climate is that Space of the Earth Contained between three Parrallels, the middle most whereof Divideth it into two Equal Parts, serving for

for the setting out the Length and Shortness of the days in every Countrey: and for as much as there have been several ways used by the Ancients in Dividing the Climates, I shall at present Content my self with this only Division; by Reckoning the Differrence of each Climate to be one Quarter of an hour, and so the Climates are 24 in Number; till you come to the Latitude of 66 Degrees, 31 Minutes, (taking up 48 Parrallels.) To which are added six Climates more, where you will find the days to be One, Two, Three, Four, Five and Six Months long at the very Pole it self, as you may plainly see in the Annexed Table; Where you may Note, that the greater the Latitudes are, the longer the days are.

A Table of Climates, *From the Equinoctial North and South to the Poles, wherein you may see in every Clime the length of the day in Hours and Minutes;* As for Example, *in the Climate or Parrallel of* 50 *Deg.* 33 *Min. you will find the longest day to be* 16 *Hours, and* 15 *Minutes, and in the Margent you find what Inhabitants dwell in those Climates* viz. *the* Amphiscii, Heteroscii, *and the* Periscii.

	Latitud.		Lon. D.			Latitud.		Lon. D.	
	D.	M.	H.	M.		D.	M.	D.	M.
	00	00	12	00		59	59	18	30
Amphi-	04	18	12	15		60	40	18	45
scii.	03	34	12	30		61	18	19	00
	12	34	12	45		61	53	19	15
	16	43	13	00		61	25	19	30
	20	53	13	15		62	53	19	45
	23	10	13	30		63	22	20	00
	27	36	13	45		63	40	20	15
	30	47	14	00		64	06	20	30
	33	45	14	15		64	30	20	45
	36	30	14	30		65	49	21	00
	39	02	14	45		65	06	21	15
	41	22	15	00		65	21	21	30
	43	32	15	15		65	35	21	45
	44	29	15	30		65	49	22	00
	47	20	15	45		65	57	22	15
	49	01	16	00		66	06	22	30
	50	33	16	15		66	14	22	45
	52	58	16	30		66	20	23	00
	53	17	16	45		66	25	23	15
	54	29	17	00		66	28	23	30
	55	34	17	15		66	30	23	45
	56	37	17	30		66	31	24	00
Hetero-	47	34	17	45					
scii.	58	26	18	00		67	15		01
	59	14	18	15		69	30	Months.	02
			Periscii.			73	20		03
						78	20		04
						84	09		05
						00	00		06

CHAP. V.

Of the Division of the Inhabitants of the Earth, Respecting their Site and Position, in Reference to One another.

1. THe Inhabitants of the Earth are divided into the *Periæci*, *Antæci*, and *Antipodes*.
2. The *Periæci* are such as dwell in the same Parallel on the same side of the Equator, and opposite to us: These live in the same Zone, and the same Clime; and cast the same shade with us: These Enjoy our Portion of heat and cold, our Seasons of the year, our Increase of Days and Nights, and all things else of this kind; Saving that our Hours are Opposite, their six in the Evening, is our six in the Morning, our Noon their Midnight.
3. The *Antæci* are Inhabitants that are under our Meridian, which makes our hours, and theirs the same; but by being 51 Deg. 30 Min. on the other side of the Equator, it happens that though we all agree in the Temperament of Zones, Number of Climes in Casting a Shadow to one side only, and the like; Yet their Zone and Climate is Southern, their Shadow falls to the South Pole, their Winter is our Summer, our Spring their Autumn.

4. The

4. The *Antipodes* are such as dwell Feet to Feet, and are in height of Oppofition, and differ in all things; our Summer is their Winter, our Noon is their Midnight; we have the North Pole Elevated, they the South. This truth of the *Antipodes* was in former time Reckoned fo Ridiculous and Impoffible, that *Virgilius* Bifhop of *Salisburgh* who writ a Treatife thereof, was Condemned of Herefy by Pope *Zachary*, in the year of our Lord, 745.

CHAP. VI.
Containing feveral Ufeful Diftinctions in Geography.

THe *Latitude* of a place is its neareft diftance from the Equator, either to the Northward, or Southward thereof, meafured in the Meridian.

2. The *Longitude* of a place is the Number of Degrees, (Reckoned Eaftwardly in the Equator,) from the Grand Meridian to the Meridian of the place required.

☞ And here Note that in all the Mapps in this Treatife, the Longitude beginneth at the Meridian of *Pico Tenariffa*.

Zenith, is a Point in the Heavens that is Right over our heads, and is fometimes called the Vertical Point, and Pole of the Horrizon.

Nadir, is a Point in the Heavens, Oppofite to it, right under our Feet.

Of

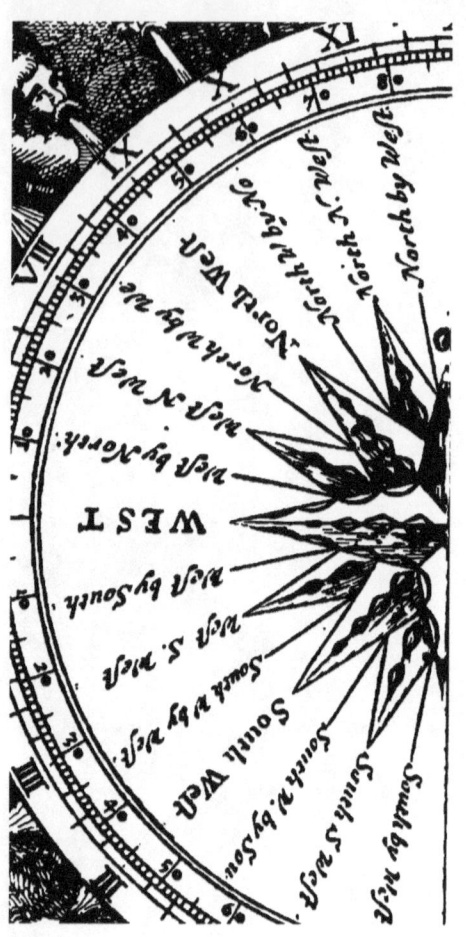

of Geography.

Of the 32 *Points of the Mariners Compass, which are thus Divided.*

The four first, are the Cardinal Points, and are Composed of one Syllable, as East, West, North, South.

The Four Seconds are Colateral Points, Consisting of two Syllables, as North East, North West, South East, South West.

The Eight Thirds are, those that are Composed of Three Syllables, as North North East, North North West, South South East, South South West, East North East, East South East, West North West, West South West.

There are Sixteen Inferiour Points, Eight of them are Composed of four Syllables, and the other Eight of five Syllables: Those of four Syllables, are these that follow.

These are Composed of four Syllables.	*These are Composed of Five Syllables.*
North and by East.	North East and by North.
East and by North.	North East and by East.
East and by South.	South East and by East.
South and by East.	South East and by South.
South and by West.	South West & by South.
West and by South.	South West and by West.
West and by North.	North West and by West.
North and by West.	North West & by North.

The Figure of the Mariners Compass is here Annexed.

Of

Of Measures.

Twelve Inches makes a Foot.
One Foot and a half, one Cubit.
Two Foot and a half, a Common Pace.
Two Common Paces, one Geometrical Pace.
Three Foot one Yard.
Six Foot one Fathom.
Sixteen Foot and a half, one Pole or Perch.
Forty Perches makes one Furlong.
Eight Furlongs, or 320 Perches, make one *English* Mile.
125 Geometrical Paces makes one Stade.
Eight Stades, or 1000 Geometrical Paces, is equal to an *Italian* Mile.
1250 Geometrical Paces is one *English* Mile.

60 *English* Miles hath Vulgarly been accounted one Degree on the Earth, but really and truly, (As hath appeared by very Worthy Experiments,) that 69 Miles and a half of our Statute Miles, makes one Degree on the Earth: But according to the Vulgar Measure, the Compass of the Globe of the Earth, is 21600 Miles, and the Diameter thereof, is 6875 Miles, and 4528 parts; which Diameter being Multiplyed by the Circumference, gives the quantity of Miles in the Superficies of the Earth and Water, And the Superficies being Multiplyed by $\frac{1}{6}$ of the Diameter, gives the Solidity in *English* Miles.

1500 Geo-

of Geography. 1

2500 Geometrical Paces make one *Scotch* Mile.
2500 Geometrical Paces make one Common *French* League.
3400 Geometrical Paces makes a *Spanish* League.
4000 Geometrical Paces makes a *German* League.
5000 Geometrical Paces makes one *Swedes* or *Swisses* League.
6000 Geometrical Paces makes one *Hungarian* League.

CHAP. VII.

Of the Use of the Mapps and Tables Contained in this Book.

THat which hath been already difcourfed, hath been in the Nature of an Introduction; to give you fome Light and Underftanding in the Principles and Elements of Geography: that Concernes the feveral Diftinctions and Divifions of the Earthly Globe, by Circles, Zones and Climates; Difcovering the various Pofitions of the Inhabitants, in Refpect of their Shadows, and Contrariety of Seafons, &c. Which may very well ferve as a good Introduction to that which follows.

Therefore take notice, that to each Mapp there is a Table, that is as an Index to fhew you what Divifions are in the Mapp, and is as it were an Explanation

planation thereoff, which you will find very useful for the understanding of them.

As for Example, In the Mapp of the World, the Table Informs you of the General Division of the World, which is thus worded, *The Mapp of the World is divided* into two Continents; The Continent of *Europe*, *Asia* and *Africa*.

The Continent of North-*America*, South-*America*.

Now if you cast your Eye upon the Mapp, you may there plainly perceive the same Divisions Circumscribed with one intire Colour with the Name in the midst thereof, in Remarkable Capital Letters.

Then if you desire to know how any of the Quarters are Divided; Pitch upon what Quarter you please, (Suppose *Europe*;) Then Apply your self to the Table, and there you will find the General Divisions of *Europe* into the *Empires* Kingdoms, Principalities and Dominions, &c. Contained in the same, with the Principal Cities in each Country; then turne to the Mapp, and you will see each Respective Division, Exactly Answer thereunto, as it is Exprest in the Table.

Now, if you would descend into the particular Divisions of any one of the Quarters, you must proceed in the same Order, as has been Directed; Suppose it were *Germany* :) Therefore turn to the Table, and there find into what parts *Germany* is Divided: as *Mecklenburgh*, *Pomeren*, *Brandenburgh*, &c. with the Principal Cities contained therein; so that by this you may note the Excellency of this Contrivance, for by these Divisions, appearing so plainly to the eye, that you may see how one Countrey border upo

An Inſtrument for finding the hour of the Day (at all times) in any part of yͤ World.

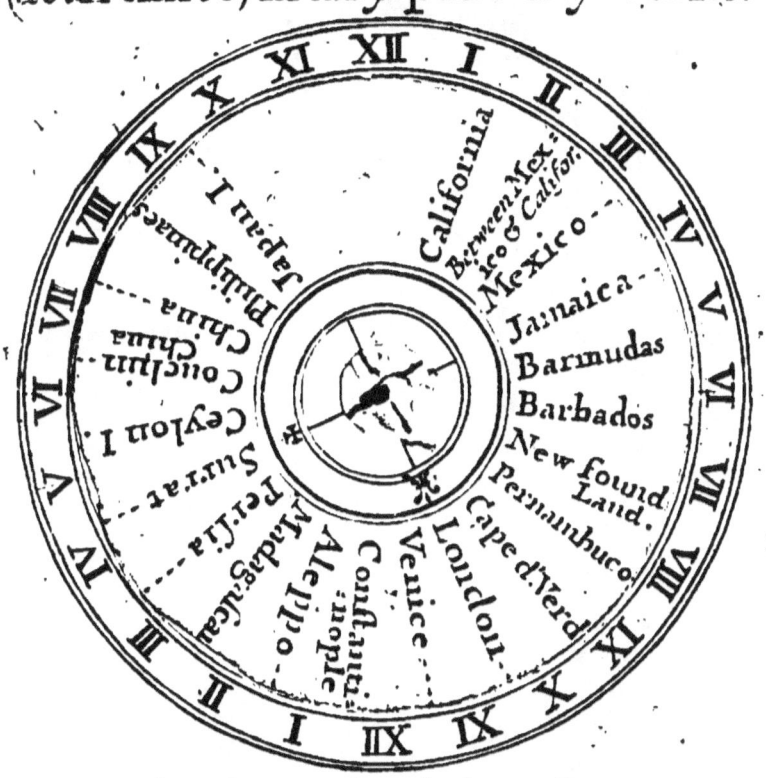

An Example of the Uſe of this Inſtrument

If it be 9 a Clock at London what hour is it at any other part of the World, therfore turn London to 9 (the hour given) Then ſhall yͤ preſent hour be ſhewn at all thoſ places Expreſt in yͤ Rundle; as you ſee it is 11, at Conſtantinople & 12 at Aleppo &c.

he WORLD
ny part of the World, And to know
are at Dinner, wher at Supper, and where
er the World.

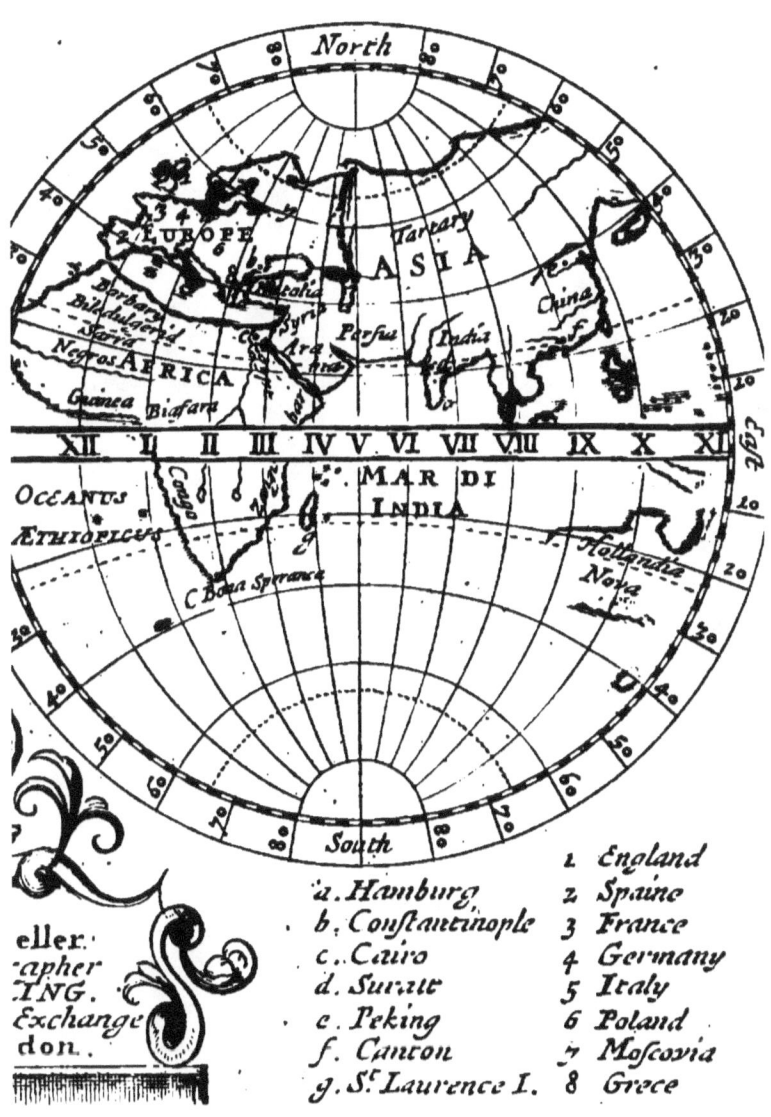

1. England
a. Hamburg   2. Spaine
b. Constantinople 3. France
c. Cairo   4. Germany
d. Surrat   5. Italy
e. Peking   6. Poland
f. Canton   7. Moscovia
g. S.t Laurence I. 8. Grece

upon another, so that by this View an Idea of the whole Country (with all its Divisions) may be framed in your mind, that at any time you have the whole Prospect thereof drawn into your Imagination, and Understanding; and can thereby Conceive the Probability or Improbability of any matter, that may be discoursed of in those Countreys.

Now, if you would yet descend into the knowledge of any one particular Countrey, (or Subdivision in this or any other Empire or Kingdom that is Exprest in the General Table,) you are to proceed in the same Order and manner as is before Directed.

CHAP. VIII.

Shewing the Use of a Mapp of the World that gives the hour of the day, (at any time,) in any part of the World; and to know where the People are Rising, where they are at Dinner, where at Supper, and where going to Bed; being an Opperation most Diverting and Pleasant, to be performed by a Map.

IT is in all Respects as other Mapps of the World, save only that this is Divided into 24 Meridians, or hour, Circles-Numbred in the Equinoctial with I. II. III. IIII. V. VI. &c. to XXIIII. the hour Circle

Circle of XII. paſſing through the Zenith of *London*.

The Uſes Follow.

To find the hour of the Day (at any time) in any part of the World.

Having the Mapp before you, you may take notice that when it is Twelve a Clock at *London*, then whereſoever you caſt your eye, it is the ſame hour at Every Place, as is Expreſt in the Mapp, *viz.* It is one a Clock at *Venice*, Two a Clock at *Conſtantinople*, Three at *Aleppo*, and Four at *Madagaſcar*, &c. In the Eaſtern Hemiſphere, and in the Weſtern Hemiſphere: Six a Clock at *Jamaica*, Three a Clock at *Califournia*, &c.

But, Suppoſe it be any other hour with us at *London*, then for a more ready finding the hour in other places: I Refer you to this Adjoining Inſtrument, the Uſe of which may be made plain by one Example: As, Suppoſe it were Eleven a Clock at *London*, then turn *London* (on the moveable plate) to the hour of Eleven in the hour Circle, then will it ſhew at that very time the preſent hour, at all thoſe places in the moveable Plate; it will be Twelve a Clock at *Venice*, One a Clock at *Conſtantinople*, and Two at *Aleppo*, &c.

So the ſame Rule will tell you, at any time, the hour of the Day in any part of the World: by turning *London* to the preſent hour there, and all the reſt will fall ſucceſſively, as in the foregoing Example.

To

of Geography.

To know by the aforesaid Mapp, where the People are Rising, and where they are at Dinner, where at Supper, and where going to Bed; and where it is Midnight in any part of the World.

This Problem is grounded upon this Hypothesis, that Six a Clock in the Morning may be taken for time of Rising, and Twelve a Clock for Dinner time, and Six a Clock in the Evening for Supper, and Ten a Clock at Night for time of going to Bed; and Twelve a Clock for Midnight.

Therefore Note, when it is Dinner time with us at *London*, then they are Rising at *Jamaica*, and at Supper at *Surrat*, and going to bed in the *Phillipina* Islands, (a little to the Eastward of *China*) and Midnight in the Pacifick Sea, and by the same Rule and Reason you may know the same things at any other hour at *London*. The forementioned Instrument doth most readily perform the same Operations by Noting what hour stands against any place in the Moveable Rundle; as suppose you turn *London* to Twelve a Clock, then you will find the hour of Rising, Supper time, of going to Bed, and Midnight, at the places before Mentioned.

There is also another Pleasant Operation to be performed by this Mapp, to know what company we have of Divers Nations to rise with us, to go to Dinner when we do, and to Sup and go to bed at the same time with us; that is to say, all those Inhabitants that dwell under our Meridian, or hour Circle, which are *French*, *Spaniards*, *Moores* and *Negroes*, all which Nations Rise, Dine, Sup, and go to bed, when we do.

C 2 *Chap.*

CHAP. IX.

Giving the Content or Quantity of the whole World in Acres, and of every Kingdom and Province thereof.

THe Globe of the World is suppofed to be one third part water and Seas: And one third part not Inhabited; And one third part inhabited, which Containeth in Acres, as followeth.

In Europe.

	Acres.
England.	29000568.
Scotland.	14000432.
Ireland.	18000000.
The Ten Spanish Provinces.	7197000.
The Seven United Provinces.	3599000.
France.	82879000.
Spain.	67000535.
Italy.	44000257.

Under Spain.

Naples.	11704000.
Lumbardy.	1640000.
Savoy.	1910000.
Piedmont.	1160000.

Under

of Geography.

Under Venice.
acres.

Trevisano.	2584000.
Verona.	480000.
Friul.	1047000.
Mantua.	480000.
Tuscany.	4785000.
Suria and Florence.	1480000.
Mercia Ancona.	1412000.
Parma.	885000.
Sicilia.	3113000.

Under Rome.

Liguria.	1415000.
Romania.	1085000.
Heturia.	540000.
Latium.	480000.
Cyprus.	1601000.
Corsica.	1395000.
Sardinia.	4089000.
Candia.	2060000.

Under the Turk.

Under Germany.

Saxony.	3484000.
Misnia.	3249000.
Turingia.	1093000.
Lusatia.	2572000.
Bavaria.	3249000.
Holsatia.	3644000.
Helvetia.	1232000.
Basil.	842000.
Sweburgh.	2109000.
Saltsburgh.	1063000.
Trier Mentz Spires. Stratsburgh and Wormes. }	4237000.
Juliers.	348000.

Cleve.

	acres
Cleve.	258000.
West-Phalia.	2300000.
Osnab.	358.
Silesia.	5706000.
Bohemia.	7024000.
Austria.	6121000.
Moravia.	4114000.
Pomerania.	3249000.
Brandenburg.	6208000.
Mecklenburg.	2107000.
Franconia.	6361000.
Tyrol.	3249000.
Carinthia.	1588000.
Stiria.	1779000.
Palatine of the Rhine.	4361000.
Wortemburg.	1223000.
Emden.	230000.
Oldenbourg.	449000.
Liege.	548000.
Cologne.	215000.
In all	93000646.
Russia.	9607000.
Volhinia.	5762000.
Mussovia.	196000.
Livonia.	34115000.
Poland.	19205000.

The Dominions of Denmark.

Denmark.	10426000.
Norway.	28492000.
Holstein.	1065000.
Ditmarsh.	337000.

The

of Geography. 21

The Dominions of *Sweden.*

Sweden.	57430000.
Finland.	7531000.
Gothia.	20936000.
Hungaria, Dalmatia, Transilvania, and all Turky in *Europe*	385367000.
Part of *Russia*, or *Muscovia*, in *Europe*, Contains.	232558000.
Part of *Muscovia* in *Asia* Contains.	128817000.
Tartary.	299110000.
Persia.	385367000.
East-Indies.	587200000.

In Africa.

Africa.	1541883000.

In America.

North part of *America.*	1152400000.
South part of *America.*	1349133000.

So that the whole Sum of the Habitable Part of the World is six Milliars, six hundred and 23 Millions, and Seventy thousand Acres.

☞ Note, that a Milliar is one Thousand Millions.

A Table shewing the bearing Distance, and Longest day, and difference of Meridians of most of the principal Cities in the World, from the Famous City of *London.*

Names of the places.	The way or Point of bearing.	Dist. in miles.	Long-est day H. M.	Differ. meridi H. M.
Alexandria in Egypt.	ſ. e. by e.	2196	14 00	1 42
Amſterdam in Holland.	e. by no.	266	16 40	0 28
Athens in Greece	ſ. e. by e.	1641	14 40	0 56
Antwerp in Brabant	Eaſt fere	248	26 28	0 24
Barwick in England	Nor. fere	267	17 24	0 2
Babylon in Chaldea	ea. ſo. ea.	2724	14 25	3 56
Bethſaida in Canaan	ſ. e. by e.	2365	14 6	2 29
Bermudas Weſt Ocean	w. ſo. w.	3409	14 10	4 56
Calieute in Eaſt-Iudia	ſ. e- by e.	5214	12 20	6 8
Calis in France	ea. by ſo.	86	16 25	0 9
Conſtantinople in Turky	ea. ſo. ea.	1547	15 15	2 24
Dublin in Ireland	n. w. by w.	296	17 15	0 26
Dantzick in Pruſia.	ea. no. ea.	961	17 5	1 44
Damaſcus Turky in Aſia	ea. ſo. ea.	2404	14 15	3 16
Edenburgh in Scotland	North	328	17 40	0 0
Epheſus in Greece	e. ſ. e.	1808	14 40	2 30
Florence in Italy	ſou. eaſt.	801	15 10	0 57
Frankford in Germany	Eaſt fere	448	16 15	0 47
Hamburgh in Germany	ea. no. ea.	538	18 0	0 56
Hieruſalem in Canaan.	ſ. e. by e.	2352	14 8	3 3
Iſleland in the N. ſea	n. n. w.	930	21 44	0 52
Joppa in Canaan	ſ. e. by e.	2938	14 6	5 0
Lisbone in Portugal	ſo. ſo. w.	985	14 45	1 0
Middleburgh in Zeland	Eaſt	295	16 30	0 20
Mentx in Germany	Eaſt	410	16 25	0 45
Milhain in Italy	ſ. e. fere.	645	15 22	0 48
Morocco in Barbary	ſo. by w.	1449	14 0	0 28
Mexico in America	w. by ſo.	6844	13 20	9 59
Naples in Italy	So. by e.	1051	14 50	0 16
Niniveh in Aſia	e. ſo. ea	2635	14 30	3 52
Paris in France	ſ. ſ. e.	215	15 57	0 20
Philippi in Macedonia.	e. ſ. e.	1395	15 10	2 10
Prague in Bohemie	Eaſt fere	700	16 15	1 14
Quinzai, the greateſt City in the world.	E. by S-	7272	13 35	11 28
Rome in Italy	ſ. e. by e.	887	15 4	1 7
Spiers in Germany	e. by ſ.	430	16 2	0 46
Strawsborough	ea. by ſ.	432	16 0	2 41
Toledo in Spain	ſo. by w.	934	14 30	5 36
Troy in Greece	e. ſ. e.	1605	15 0	2 75
Venice in Italy	e. ſ. e.	744	15 28	1 3
Sivil in Spain	ſo. by w.	950	14 40	0 52
York in England	No. fere	150	17 0	0 4
LONDON.			16 26	0 0

FINIS

A Geographical Description OF THE WORLD.

CHAP. I.

Of the World in General

THe furfare of the Earth is Divided into two great Continents one of which is Divided into *Europe, Asia and Africa*, and the other into two, *viz.* North and South *America*, as are plainly seen in the Mapp of the World, with their names in the midst in Capital Letters: There is but little difficulty about the bounds after that which joyns *Europe* and *Asia*, is Considered (for bating the little *Isthmus* made by the Mediterranean and *Mare Rubrum*, which containes the Limits of *Asia* and *Africa*) there is no part of the said Quarters which is not Surrounded by the Sea: For the separation then or Boundary between *Europe* and *Asia*, Geographers are of various Opinions Concerning this Division; Some Divide it thus, with a line drawn through the *Egean* Sea and *Helespont*, through the *Euxine* Sea to *Palus Meotis*, along the stream of the River *Tanais* to the mouth thereof, and so by the River *Dwina* to the Bay of St. *Nicholas* in the white Sea.

Others (as the Right Honourable the Earl of *Castlemain*) in his Book of the use of the English Globe, doth more Judiciously divide it thus, Supposing a line drawn from the Mouth of *Tanais* Runs up the stream to *Tuia* (Scituated on the most Easterly flexure,) and thence going to the nearest Banks of the River *Oby*, accompanies it till it falls in the Northern *Ocean*, my Lord doth the rather Incline to this Division than any other, because it Containes almost all the Vast Dominions of the Russian Emperour, and so makes him an intire *European Monarch.n*

CHAP. II.

Of Europe in General.

EUrope although it be the least of the four grand Divisions of the Earth (as appears in the Mapp of the World) yet it is of the most *Renowne*.
1. For the *Temperature* of the Air, lying in the
midst

midſt of the *Temperate* Zone, and therefore Fertile in its ſoil. 2. The flouriſhing of Arts and Sciences. 3. For the Purity and Sincerity of the Chriſtian faith,

The language of the *Europeans* is Divided into ſeveral branches or *Dialects*, but all from three Roots or ſtems, which are, 1. The *Latine*. 2. The *Tutonick* or *High Dutch*. 3. The *Sclavonian* tongue. Thoſe that Branch from the *Latine* are the *Italians*, *French*, *Spaniards*, *Portugalls*; and thoſe from the *Tutonick*, are the *Engliſh*, *Dutch*, *Swedes*, *Danes*, *Gothes*; and from the *Sclavonian*, the *Croatians*, *Hungarians*, *Tranſilvanians*, *Ruſſians*,

The Kingdoms and Principal Regions are as followeth (as they ſtand in the Table of *Europe*,) viz. *England*, *Scotland*, *Ireland*, *Spain*, *Portugal*, *France*, *Italy*, *Germany*, *XVII Provinces*, *Norway*, *Sweden*, *Denmark*, *Poland*, *Lithuania*, *Moſcovia Ruſſia*, leſſer *Tartary*, *Turky* in *Europe*.

The moſt famous Rivers in *Europe* are nine, The *Thames* in *England*, *Tornia* in *Scandinaria*, *Wolga* in *Muſcovia*, the *Loire* in *France*, the *Rhine* in *Germany*, the *Weyſſel* in *Poland*, *Tagus* in *Spain*, *Po* in *Italy*, and the *Danube* in *Germany* and *Turkey* in *Europe*.

We Rank *England*, *Scotland* and *Ireland*, in the firſt place, in Regard they owe obedience to the Scepter of our Gracious Soveraign.

We ſhall therefore Begin with the Deſcription of the Kingdoms of *Great Britain* and *Ireland*, (not affecting that *Diminutive Appellation* of the Britiſh Iſles) as if they were *Guernſey* or *Jarſey* and no way Comporting with the Renown and Grandure of thoſe famous Kingdomes.

Of

Of Great Britain.

GReat Britain Containes *England*, *Scotland*, and *Wales*, making the moſt Famous Iſland in the whole World. It was once called *Albion, ab Albis Rupibus* from the white Rocks apearing on the South Coaſt, towards *France*, from whence it was firſt *Diſcovered*. Afterwards *Britain*, ſome ſay from *Brutus*, of the Trojan Race, who firſt ſetled a Government here; or as others will have it, from *Prutainia*, ſignifying Mettals, becauſe of the great quantity of Tin, Iron, Lead, &c. that is found here. But others ſay the Britaines had their name from the word, *Brith*, which ſignifieth ſtained or painted, by reaſon of a Cuſtome the Natives had to paint their Skins all over, and of ſeveral Colours, that they might thereby the more Terrify their Enemies, whence alſo the Romans called the People of *Scotland Piſts*.

The whole length from *Strathy*-head in *Scotland* to the *Lizard*-point in *Cornwall*, is counted 624 Miles: And the breadth from the lands end in *Cornwall*, to the Iſle of *Thanet* in *Kent* is about 340. It is obſerved in his Majeſties Teritoryes about *Great Britain* and *Ireland*, there are eight Several *Dialects* ſpoken by his Majeſties own Subjects, *viz.* 1. *Engliſh*. 2. *Scotch*. 3. *Iriſh*. 4. *Welch*. 5. *Corniſh* (in Cornwall.) 6. *French* (in *Garnſey* and *Jarſey*.) 7. *Manks* (in the Iſle of Man) and 8. *Gothiſh* (in the Iſlands of Suhtland.

Of *England*.

ENgland (a name taken from the *Angli* a people that came in with the Saxons, and not chang'd by the Danish or Norman *Conquerors*) is the chief part of the Island of Great Britain, being Divided into 40 Countys and 22 Bishopricks, is Accounted in length 386 miles, in breadth 279. The soil is very fertile and plentifull; several of its Chief Commodities and Excellencies are Contained in this verse,

Anglia, *Mons*, *Pons*, *Fons*,
Ecclesia, *Femina*, *Lana*.

England is stor'd with Bridges, Hills and Wooll, With Churches, Wells, and Women Beautifull.

Its first known inhabitants were the *Britains*, who being Conquered by the *Romans*, and afterwards over-run by the *Saxons*, were forced to Retire into that Corner of the Kingdom called *Wales*: where their Posterity to this day Inhabite, being a Province Divided into 12 Countys and 4 Bishopricks, the eldest Son of our English Kings, being always Entituled Prince of *Wales*.

The forementioned *Saxons* Divided the whole Realm into seven Kingdoms, and were much oppressed for a long time by the *Danes*, but at last

28 *A Geographical Description*

laſt being united under one King, were ſubdued by the *Normans*, under *William* the *Conqueror*, whoſe Succeſſors Continue to this day.

Of *Wales*.

Wales is Invironed on all ſides by the Sea, except towards *England*, from which it is ſeparated from the RIVER DEE, and a line drawn to the RIVER WYE: But Anciently it Extended to the RIVER SEVERNE Eaſtward, for *Offa* King of the *Mercians*, forced them to leave the Plain Countrys beyond that River (which now is called the *Marches of Wales*,) and to betake themſelves to the Mountaines; which he cauſed to be ſeparated from *England*, by a great Ditch called *Offa's Dike*, in *Welch Claudh-Offa*, in many places yet to be ſeen.

The whole Country is Generally Mountanous and Barren, yet affordeth ſeveral good Commodities, and is not without many fertile Valleys. which bear good Corn, and breedeth abundance of ſmall Cattle, with which they furniſh *England*; as alſo with Butter, Cheeſe, woolen Cloth, called *Friezes Cottons*, Bays, Calveskins, Hides, Honey, Wax.

It is divided into 4 Circuits for the Adminiſtration of Juſtice, and is divided into 13 Counties, wherein is contained 1016 Pariſh Churches, 56 Market Towns and 40 *Caſtles*, 230 Rivers, 99 Bridges, 32 Parks, 13 Forreſts, 1 Chaſe. Alſo theſe hills are famous for their height, *viz. Snowden, Plinillimon, Brechin, Moyluodian, Brethen, Caddoridrie, Rardiivaire, Monitch, Dennye,* and *Penman-Maur.*

Of

Of Scotland.

Scotland is the Northern Part of *Great Britain*, longer then *England*, but not so broad, much colder and less Fruitfull, the whole Containing 35 Shires, is Divided into Lowlands, which speak a kind of Barbarous English, and the High-lands, whose language is very neer the same with the *Irish*. This Realm, upon succession of King *James* to the English Crown, became united to that Scepter, and are Joyntly stiled *Great Britain*. Its Chief Commodities are Course Clothes, Frieses, Fish, Hides, Lead-Ore, and here are four Universities, *Edenbrugh*, *Glascow*, *St. Andrews, and Aberdeen*, two Arch-Bishops and eleven Bishops.

As to their Courts of Justice, they are peculiar to themselves, the chief of which is the Session or *Colledge* of Justice, onsisting of a President, 14 Senators, 7 of the Clergy and as many of the Laity, unto which was afterward joined the *Chancellor*, who is chief, and 5 other Senators, & in time of the Sessions of Parliament there is a high *Commissioner* constituted by his Majesty to Act as *Vice Roy* under him, which at present is the Illustrious Prince *James* Duke of *York* and *Albany*.

Of IRELAND.

Ireland is the bigest Island in *Europe* after Britain. The West of which lyes towards *Spain*, Containing in length above 300 Miles, in breadth 150. It neither breeds nor will Harbour venomous Creatures; the Soil is very good, and by the late Industry of the Inhabitants (now the greatest part English;) the whole Country begins to Grow rich and Flourishing, as the Populous and fair new buildings at *Dublin*,

and

and other Places, Demonstrate. It is Divided in four Provinces, *viz. Leinster, Ulster, Munster & Counaught*, and each of them into several Countyes, as may appear in the Table of *Ireland*.

The Comodities are store of Cattel, as also Tallow, Butter, Cheese, and Wool, of which they make cloth, Frieses, Ruggs, Mantles, &c. Its Seas yeilds great plenty of Codfish, Herrings, Pilchers, &c. and in the Bowels of the Earth, Mines of Tin, Lead, and Iron.

Of *Spain*.

Spain, the most Western Country of *Europe* is separated from *France* on the Northeast by the *Pyrenean Mountaines*, and on the West joyning to *Portugal*, on all other parts washt by the Sea, its whole Circuit being 1900 English Miles. They are a mixt People, Descended from *Gothes*, *Saracens*, and *Jewes*: from the *Jewes* they Inherite Superstition, from the *Saracens* Melancholy, from the *Gothes* desire of Liberty. They are much given to Women, vain glorious, and stately; very Grave in their Carriage, zealous Adherers to the Church of *Rome*, Obedient to their Prince, true to their words, and of Wonderfull Patience, Enduring adversity: their Women sober, loving to their Husbands, and Gallants, wonderful, Delicate, Curious in Painting, and perfuming, but by Custome forbidden to drink any Wine, at least till they are Married. This Country yields Sacks, Sugars, Oyl, Metals, Liquorish, Rice, Silk, Wool, Corke, Rosin, Steel, Oranges and Lemmons, and Raisins, &c. But is defective in Corn, and are glad to be supplyed therewith from *Italy*, *France* or *Sicilia*; nor are their Cattle large or many, their fare being most upon

Sallats

Sallats and Fruits; every Gentleman limitted what flesh he shall buy for himself and Family: they have Univerfities, such as they are: their Courts are kept at *Madrid*.

Of *Portugal*.

POrtugal is a Diftinct Kingdome by it felf, though anciently a province of *Spain*. Its Chief City is *Lisbon*; their Cuftomes and Religion much like the Spaniards.

Of *France*.

THe Kingdom of *France* is for one intire thing, one of the moft rich and abfolute *Monarchies* of the World, being almoft a fquare, each fide containing about 600 Miles; it is bounded on the North with Lower Germany; on the North-Weft, wafht with the Britifh *Ocean*, on the weft with the *Aquitain* Sea, on the South the *Pyrenean Mountains* fence it from Spain, on the Eaft it extends to the *Medeteranean*, and the *Alps*, which Divide it from *Italy*. It abounds with all manner of good Fruits, befides great ftore of Fifh and Fowl; but its Chiefeft Commodities are Wine, Salt, Linnen, Cloth and Corn; the lefs Materiall are Wood for

C   dying

dying, Paper, Nutts, Almonds, Laces to the New Modes. The People are general Complemental, inconstant and Rash; both their Gentlemen and Citizens live more sparingly then the English, they feed most on Boyl'd or Liquid Meats, and are very curious in sawces. In Matters of Warr, there is an old Comparison that Resembles the *French* to a Flea, quickly Skiping into a Country, and assoon leaping out of it; of which late Transactions afford several instances; their Language is *Amorous*, and they leave out manny *Consonants* in *Pronuntiation*, Occasioning this *Proverb*, A Frenchman writes not as he pronounces, Sings not as he Pricks, nor Speakes as he thinks. The Nationall Religion is Popery, yet there are many Protestants amongst them, whom they in derision call *Hugonots*, who of late have bin and are under great Persecution for their *Religion* and loss of their Priviledges, not occasioned in the least by their disloyalty, or any disturbance of the Government, but only and alone from the Malice of the Popish Clergy. The Air in the Northern parts of *France* is Purer then that of *England*, and being not so much Covered with Clouds drawn out of the Sea, becomes more cold in Winter, and more hot in Summer; and less Annoyed with mists and Rain.

The Kingdom is divided into twelve General Governments, to which may be added four new Governments, being the late Conquests of the *French King*.

Four towards the North on this side the *Loir*, four in the middle of *France*, on each side the Loir, four on the South beyond the *Loir*, and the other four to the Eastward towards *Germany*.

Of

Of several Countrys bordering upon *France*

LA Franche Comte, or the free County, called also the County of *Burgundy*, is invironed with *Champaigne*, *Lorrain*, *Breſſy*, and the *Dutchy* of *Burgundy*, being in length 90 Miles, and breadth 60. One of the moſt fertile Provinces in the World, the chief City is *Beſanſon*, the next *Dole*, famous for the Colledg of *Jeſuits* there. It was under the *Spaniſh* Crown, but a few years agoe ſudenly ſurprized by the *French*.

2. *Lorrain*, a Principality adjoyning to that laſt mentioned, about four days Journey in length, and three in breadth; it abounds with Corn and Wine, good Horſes, plenty of Mines and Store of Salt and Fiſh: the Dukes Revenue was formerly computed at 700000 Crowns *per Annum*; and they were wont to give for their Device, An armed Arm coming as it were from Heaven, and Graſping a naked Sword, to ſhew that they were abſolute Princes, holding theire eſtate by no other tenure then from God and the Sword: but the *French* have likewiſe not long ſince violently over-run; this Countrey; and the preſent Duke Commands the *Emperours* forces, having Married the Queen *Dowager* of *Poland*.

3. *Savoy*

3. *Savoy*, a Dukedome compassed with *Dauphine*, *Switzerland* and *Piedmont*, which latter is for the most part under its Government, lying on the *Italian* side of the *Alps*, and being exceeding pleasant and Fruitfull. The rest of *Savoy* is *Mountanous* & *Barren*. Adjoyning hereunto, on the Lake *Lemane*, stands the City of *Geneva*, a Common wealth, not extending eight Leagues in Circuit; But of indifferent good Trade, and well Fortified.

4. *Switzerland*, is likewise a Republick, Consisting of 13. Shires or *Cantons*, of which five are all *Papists*, six all Protestants, and two mixt. 'Tis thought to be the highest Countrey in *Europe*, and sends forth four of the Greatest Rivers, *viz.* The *Danubius* through *Germany* and *Turkie* East; the *Rhine*, through *Germany*, North; the *Rhosne*, through *France*, West; and so through *Italy*, South; the Soil is but mean, being so Hilly; but the Men great Warriours, and famous for their Mercinary Valour.

This Country is in length about 240 miles, and 188 in breadth, very Mountanous, affording Deers, Wild-Goats and Bears. The Valleys affords rich Pasture for Cattle, wherein consists their greatest wealth, in some places they have good Wines and Corn.

Of

Of Italy.

Italy is the moſt famous *Region* of *Europe*, very much Reſembling the form of a mans legg; A moſt Pleaſant Countrey, Commodious for Traffique, and Exceeding fertile.

'Tis bounded on the Eaſt with the *Adriatick* Sea, South and Weſt with the *Tyrrhene* Sea, on the North with the *Alps*, being in length 1020 Miles, and in Breadth not above 440. in the Broadeſt place.

It abounds with Corne, Wine and Oyl, yields plenty of Almonds, Pomegranats, &c. The whole Countrey ſeeming as an intire Garden.

The People were Anciently famous for their Valour and Prudence, and are ſtill Courteous and Grave, and very Obliging to Strangers, yet much Enclining to Jealouſy and Wantonneſs, and ſharp Revenge; ſparing in Diet, but neat; their greateſt Expences are about their Gardens.

Here about 1600 year agoe the *Latine* tongue Floriſh'd, being vulgarly ſpoken; but afterwards by the Invaſions of the *Gothes* and *Vandales*, and other Barbarous Nations, the Common Speech became mixed and degenerated, which at this day we call the *Italian*, and is ſtill a moſt Delicate, Noble, and Courtly Language.

The Capitall *City* is *Rome*, once Miſtris of the World in *Temporalls*, and ſtill pretending to be ſo in Spiritualls, in Compaſs about 11 Miles, but full of Gardens, and waſte ground, ſo that its ſuppoſed not to Contain above 250000 Soules, two parts in three of whom are Clergy men, and Curtezans.

But in the Flouriſhing of the *Roman Empire*, it contained 50 Miles in Compaſs, and not Fewer then 3 or 4 Millions of People, as is found in *Authentick Hiſtory*.

After the *Emperours* removed their Seat to *Conſtantinople*, the Biſhop of *Rome* taking Advantage by their abſence, by pretended *Donation* from *Conſtantine* made himſelf abſolute *Prince* of a great part of *Italy*, called *St. Peters Patrimony*, and the Lands of the Church, and the reſt in time became *Cantonized* into ſeverall *Petty Principalities* and States.

In the Bottom of the *Adriatick*, now called the *Gulf* of *Venice*, ſtands the famous City and Virgin Commonwealth of *Venice*, Situate on many *Iſlands*; and the water running through every ſtreet, being in compaſs 8 Miles, Containing 70 Pariſhes, many Excellent *Pallaces*, and *Curious* Buildings.

Of the Principal Iſlands in the Mediteranean Sea.

Of Sicilia.

THis Iſland is the chiefeſt of all the others in the Mediteranean Sea, and is a Kingdom of of it self, it is in Circuit about 780 Miles, it is of a *Triangular* Shape.

It is very plentiful, and abundant in most things, especially Corn; it yields most Delicate Wines, sweet Oyl, Sugar, Honey, Silk, Safron, great store of Salt, and by reason of the sweet *Temperature* they have all Sorts of Fruits, as also Mines of Gold, Silver, Iron and Allom: there are also found *Emrraulds* and *Agates*, and other Precious Stones; The Countrey breeds an Excellent Race of Horses; and amongst the Hills and Mountaines, there is the famous Mount *Etna*, now called *Mount Gibello*, although it lyes covered with Snow, yet from the topp Issues forth flame and smoke, and sometimes casteth forth such a quant'ty of Ashes, that the fields are Covered therewith.

There are at present 12 Episcopall Cityes, the three whereof, as *Palermo*, *Messina* and *Monwale*, are Arch Bishopricks, all very rich; *Palermo* is the Royal seat of the whole Kingdome; The *Sicilians* are very Ingenious and sharpe witted People, Eloquent and Pleasant, desirous of Honour; the Island is under the Power of the King of *Spain*.

Of *Sardinia*.

THe Island of *Sardinia* lyeth in the *Mediterranean* or *Tyrrhenean* Sea, whose Circuit is 560 Miles, it is Divided into two parts, one is more Hilly then the other; the Island produceth

Excellent Wines, and abundance of Cattell, and great store of Cheese; they have good Horses, but not very high: here are many hot Bathes, Salt pitts, Mines of Silver, Brimstone and Allom. This Island hath two Arch Bishopricks, the one is *Cagliari*, and is the chief City, Seated on a Mountain, having a fair Port; the other is *Oristano*, very meanly Inhabited by reason of a bad Ayre; is has a Port, and a great River. The *Sardes* are a Rude People, and uncivil, well able to endure Labour and hardness, good Hunters, not daynty in food, not Curious in Apparell. They formerly had a language of their own but, now Corrupted, and in the the Cities they speak *Spanish*.

Of *Corsica*.

THis Island hath been Inhabited with divers Sorts of People, it is in length 120 Miles. It is Divided into two parts, the Easterne part is called the *Interior*, and the other on the West called the *Exterior* side; the Island is every where Enclosed with *Rocky Clifts*, and within the land very Hilly; it produceth Horses of a fierce Nature, and Hounds of a wonderfull Bigness.

First this Island was possessed by the *Tyrheans*, and afterwards by the *Carthaginians*; but they were driven out by the *Romans*, which were also Expulsed by the *Saracens*, which were likewise subdued by the *Genoëes*, under whose Jurisdiction it remaines.

Of *Malta*.

THis Island is renowned for the residence of the great Master of the Knights of *Jerusalem*, the Circuit is 60 Miles, the land is Stony without River in it; they have store of Sheep, Goats, Oxen, Asses, Mules, Coneyes and Partridges; the Inhabitants are very brown, swarthy Colour, by reason of the heat of Summer; the Women are fair, but they hate Company and when they goe abroad they are Covered. Upon this Island was the Shipwrack of *St. Paul*, the chief town and seat of the Bishop is called *Malta*.

Of *Corfu*.

COrfu is one of the Islands in the *Ionian* Sea, in Circuit 300 Miles, and 40. in length; the South part Hilly, but the North part plain, one *Mountain* Excepted, which stands neer the Sea, on whose top is seated *Castle Nova*, underneath it *Castle Vechio*, and at the foot of the Hill lyes the City of *Corfu*, shut in and enclosed between those two Castles; the Aire is very sweet and good; it abounds with Cedar trees, Orange trees, and other Fruits; it is Plentifull in Honey, Wax, Wine

and

and Oyl, Fish and wild Fowl, and wild Beasts, to the great pleasure and pastime of the Inhabitants in their Hunting and Hawking Recreations.

Of *Candia*.

THis Island is one of the most famous in the *Mediterranean* Sea, it reaches 270 Miles in length, and 50 Miles in breadth. This Isle was ormerly called *Crete*, and had at that time 100 Cities; it is Hilly in the Easterne part: it abounds with Olive trees, Oranges, Cedars, and Excellent Wine, called *Malmsey* or *Muscadine*, and in their Valleyes Exceeding fair Pastures. This Island hath been lately famous for the great and eminent defence it made against the *Turk*, but at last was forc'd to yield the Conquest thereof to them, so the *Venetians* lost it, after a long and tedious siege. The Chiefe City of the Island is called *Candia*.

Of *Cyprus*.

THis Island is 550 Miles in Circumference, in length it is 200, and in breadth 65 Miles. It is Divided into two parts by a *Mountain* which runs through it from East and West; the Ayre is very hot, and but little water, there falls little raine; it

abounds

abounds with all things needfull for life; it produceth great store of Corn, and other Pulse and Grain, Oyl, excellent Wine, Sugar, and Cotten-Wool, Honey, Turpentine, Verdigreace, Brass, and all Sort of Mettals, Salt and Grogrames of Goats hayre; the Women Lascivious.

This Kingdome is Divided into 11 Provinces, the Chief City is *Nicosia*, adorned with fair Churches and sumptuous Pallaces; it was once in the Possession of the *Venetians*, but now under the *Turk*.

Of *Majorca* and *Minorca*.

TWo Islands about 60 miles from *Spain*, the first 300, the second is 150 Miles in Circuit, and both Indifferent Fruitfull.

Of *Cephalonia* and *Zant*

THese Islands each of them are in Compass 60 Miles or there abouts, they have plenty of Currans and Oyl, and Wine; and are under the *Venetians*.

Of

Of *Germany*.

Ermany is Divided into two parts, High and Low. High *Germany* is bounded on the West, with *France* and *Belgium*; on the North, with *Denmark* and the *Baltique* Sea; on the East, with *Prussia*, *Poland*, and *Hungary*; and on the South with the *Alps* and *Italy*. The Country is almost Square, each side being 650 Miles; it hath one General Governour called the Emperour, a Name derived from the Ancient *Roman* Emperours, but retains very little of that Power; for most of the other Princes are absolute in their Respective Dominions, whence this mighty Body, by Reason of Various Interests and Differences, becomes nothing so formidable as it would be, if firmly united under one Soveraign Monarch; The Principal Regions are presented to you in the Table of *Germany*, the whole Country abounds with most things necessary for Life.

Its Commodities are Mines of Silver, an other Inferiour Metals; Wines, and fresh Fish, Quick-Silver, Allom, &c. The People are Honest, Laborious, and Sincere, Stout Drinkers, and Good Souldiers: The Women of Indifferent Complexions, but somewhat corpulent; as for their Diet, There is this Proverb, The *Germans* have much meat, but Sluttishly drest; The *French* little, but neatly Cookt; The *Spaniards* neither one nor the other.

The Title of the Father defcends to all the Children, every Son of a Duke, is a Duke, every Daughter a *Dutchefs*; For Religion, fome are Papifts, others Proteftants; which latter are again divided into *Lutherans* and *Calvinifts*.

Of the Seventeen Provinces.

LOwer *Germany* is that which is called *Belgium*, or the *Netherlands*, and is bounded on the Eaft, with the River *Ems*, and part of High *Germany*, on the Weft with the *German* Sea, on the North with Eaft *Freefland*, and on the South with the *Some*, *Champaign* and *Lorrain*, containing in all Seventeen Provinces: But we fhall only fpeak here of the feven United Provinces, or *Dutch* Common-Wealth; The Names of which are, *Holland*, *Zealand*, *Utrecht*; *Over-Iffel*, *Zutphen*, *Groningen*, and *Weft-Friefland*, which was firft made in the year 1581. on their Revolt from their Ancient Soveraign the King of *Spain*, againft whom by the great Affiftance of the *Englifh*, and Princes of *Orange*, they Waged War many years fo fuccefsfully, that he was at laft forced to treat with them as a Free State.

Thefe Countries are exceeding Populous, being a People very Induftrious, they have attempted to Grafp the Trade of this part of the World into their hands.

In thefe Provinces the Women govern all, both within doors, and without, and make all bargains, which

which makes them injurious, and Troublesom; the Eldest Daughter is of the greatest Reputation, yet hath no larger Portion then the rest; they Marry Noble with Ignoble, young with old, Master with Servants, and into strange Townes, and Forreign Countrys.

There are within *Holland* also a great Number of Lakes and standing Waters. They want both Corn, Wine, Oyl, Wood, Wool, Hemp, Flax, and almost all other Commodities; and yet there is not any Countrey in the North which abounds so much as *Holland* and *Zealand*, in almost all the forementioned Commodities; so great is the Advantages these Provinces receive by the Sea.

The ten Catholick Provinces, or Spanish Netherlands.

THe ten *Spanish* Provinces are these, *Flanders, Limburg, Luxenburg, Brabant,* the Marquesate of the Holy Empire; *Artois, Henault, Namurs, Meckline,* and *Gelderland.*

The Soil is very Fruitful, and mightily Peopled, but exceedingly wasted and impoverished by the late Wars and Incursions made in their Country by the *French* King.

In these Countrys belonging to the King of *Spain,* are 208. Walled Townes, 150. Townes priviledged, as walled Townes; 6300. Villages with

with Parish Churches, besides a great Number of Villages and Hamblets.

There are five principle Rivers in the seventeen Provinces; The *Rhine*, the *Meuse*, the *Scheld*, the *Haa*, and *Emes*.

The *Meuse* affords abundance of *Sturgeon*, so big, that some of them weigh four hundred, and some four hundred and twenty pounds, and are twelve foot long, of *Antwerp* Measure; The *Scheld* doth also abound with most sorts of Fish.

They have divers Forrests and Woods, well stored, with Red and Fallow Deer, Goates, Wild Boares, Hares, Coneys, Badgers, Wolves, Foxes, &c. which yeild good Furrs.

The Fowl in these Countries are Partridges, Feasants, Turtle-Doves, Quailes, and any sort of Birds as we have in *England*.

The two Principal Cities in these Countreys are *Amsterdam* in the Seven United Provinces, and *Antwerp* in the Ten Provinces.

Amsterdam standeth upon the Gulf *Tye*, and is built upon Piles under the Ground, as *Venice* is; so that the buildings under ground, are more chargable then above: It is very rich in Merchandise, the buildings are very Sumptuous and Fair.

Antwerp lyes on the right side of the *Scheld*, it flourisheth exceedingly in the Number of Inhabitants, in beautiful Buildings, and in Merchandizing, which is chiefly occasioned by the the Commodiousness of the *Scheld*, it being a River of so great bredth and depth, and Ebbing and Flowing so far into the Sea.

Of

Of Norway.

Norway the Western parts of *Scandanavia*, is a place very Barren and Mountainous, so that instead of bread the People eat dried Fish, which we call Stock-fish, They export Furs, Train-Oyl, Timber, Deals, Clapboard. It is a Kingdom of it self, and divided into five Governments or Provinces.

1. The Government of *Bahus*, the chief Towns, *Bahus* and *Maerstrand*.

2. That of *Agger*, the chief Towns is *Agger*, *Opslo*, and *Frederick Stadt*.

3. That of *Bergen*, the chief Town *Bergen*, the Residence of the *Vice-Roy*.

4. The Government of *Dronthem*, the chief Town *Dronthem*, the Seat of the Ancient Kings of *Norway*.

5. *Wardhuyse*, the chief Town *Wardhuyse*, near the North Cape of *Norway*.

The *Normegians* are little given to sickness, and are of a strong Constitution; their great inclination to Sorcery, makes them have the Reputation of selling the Winds to the Seamen.

Of Sweden.

THe Kingdom of *Sweden* hath on the East *Muscovia*, On the West the *Dofriae*-Hills (that part that side from *Norway*) on the North the Frozen Seas, and on the South the *Baltick*. The Country is little less then *Italy* and *France* put together. The People are good VVarriers, and live hardly, the Ayr is sharp but so salubrious, that it is ordinary for men to live 100 years; It aboundeth with Lead, Iron, Copper, Tarr, Furrs.

Sweden Comprehends seven parts, which are.

1. *Halland, Schonen* and *Bleking*, all which goe under the name of South *Gothland*. The Chief Towns *Lunden, Malmugen, Landskroon*, and *Christianstadt.*

2. The North part of *Gothland* is divided into *Ostro Gothland* and *Westro Gothland*; The Chief Towns *Calmar, Gottenburgh*, and *Linkopen.*

3. *Sweden* containing eight or nine Provinces. The Chief Cities are *Stockholme, Upsal* and *Nikopen.*

4. *Lapland* is Divided into five Regions or Territories, *viz. Uma, Piiha, Lula, Tormia* and *Lapmark*. with the Towns of the same name whereof *Tormia* is of most importance.

5. The great Province of *Finland*, subdivided into several small Provinces, The Chief Towns are *Abbo, Wiburg* and *Roseburg.*

6. *Ingria*, The Chief Towns, *Noteburg* and *Juanisgrod.*

7. *Li-*

7. *Livania*, or *Liffland*, The Chief Town, *Riga*, *Derpt*, and *Revel*.

The Crown of *Swedland* is also possessed of several Islands in the *Baltick* Sea; And in *Germany*, the Dukedoms of *Pomeren*, *Bremen* and *Ferden*; The Town of *Wismar*, and the Island of *Rugen*.

In the North Part of *Sweden*, *Tormia* and *Kimi*, are the most Considerable Rivers of *Scandinavia*.

The Country is full of Mountains and Woods, The Commodities of the Country, are Copper, Butter, Tallow, Hides, Skins, Pitch, Rosin, Timber and Boards.

There are so few sick people among them, that Physitians and Apothecaries have little or no Practice.

The Air is so sharp and salubrious, that it is ordinary for men to live an hundred years. Their Religion *Lutheran*

Of Denmark.

Denmaark Consists of three Parts, to wit, *Jutland*, which is a great *Peninsula*, or *Chersonesus*, annexed to *Germany*.

Jutland is Divided into North and South *Jutland*.

1. North *Jutland* includs four *Lutheran* Dioceses, which are,

1. The Diocess of *Rypen*. 2. Of *Arhusen*. 3. Of *Wiburg*.

Wiburg, and *Alburg*, and *Schagen*, the North Cape of *Denmark*.

2. South-*Jutland* Comprehends two Dukedoms.

1. The Dukedom of *Holstein*, Included within the Circle of Lower *Saxony*.

2. The Dukedom of *Sleswick*, the Chief Town of the same name, with the Castles of *Gottorp*, *Tonningen* and *Fiendsburg*.

The most Remarkable Islands of the *Baltick* are.

1. *Zeland*, the Chief City *Copenhagen* the Residence of the King, whose Brother Prince *George* was Marryed to the Illustrious Lady ANN, Daughter to his Royal Highness the DUKE of *York*, *Anno* 1683.

2. The Island *Funen* the Chief Towns *Odensee*, *Middlefort* and *Niburg*.

The Islands and Countries in the Northern Sea, which depends upon the Kingdom of *Denmark*, are a great Island of *Iselandia*, the Chief Town, *Hola* a Bishoprick.

The Islands *Fezo*.

The Kingdom of *Norway*, &c.

Of Poland.

THe Elective Kingdom of *Poland*, lies to the Eastward of *Germany*, on the North of *Hungary*, *Transilvania*, and *Moldavia*, and South-westward from *Muscovia*.

It Contains two Estates, that of true *Poland* with

the Provinces Annexed thereto, and the great *Dutchy* of *Lithuania*, with its Dependancies, which is now united to the Crown of *Poland*.

The whole Kingdom is divided into twelve Provinces as may appear in the Table. It is in Compass 2600 miles, very fruitful in Barley and Cattle, & Hemp Wax, Pitch & Tarr. and other Merchandize. The Inhabitants Excellent Souldiers; They are all pestered with factions, that they can attempt no great matter, only to defend themselves from the *Turks*. They are much addicted to the *Latine* Tongue.

They have of all Religions amongst them, but the Present King seems much to favour Popery; and was Eminently serviceable in the late War in the relief of *Vienna*, and the regaining of *Hungary* from the *Turk*.

Of *Lithuania*.

Lithuania is the Greatest Province of all those which Compose the Estates of the Crown of *Poland*; It has the Tittle of a Grand Dukedome, wherein there are as many great Officers, as in the Kingdome of *Poland*.

The Country is full of Marshes and Sloughs, that there is no travelling in the Winter for Ice.

Samogitia, a Countrey in this Dukdome, whose Inhabitants live very poorely.

Volhinia, the Chief City *Kiou*, an Ancient place, having once 300 fair Churches, but destroyed by the *Tartars*; still a Bishops See, acknowledging the Partriarch of *Mosco*, and of the Communion of the Greek-

Greek Church; feated on the *Boryſthenes*, where the *Coſſacks* have often had their Retreats. It was lately in the hands of the *Muſcovite*, but now ſaid to be the *Turks*, in the year 1678.

In *Podolia* ſtands the well fortified and Impregnable *Caminiack*, which formerly withſtood the Armyes of the *Turkes*, the leſſer *Tartars*, the *Tranſylvanians*, and the *Walachians*; but at length was forced to yield to the Grand Seignior, in the year 1672. ſince Retaken by the *Poles*; but by the laſt Treaty delivered to the *Turks*, as alſo *Orʒanthow* and *Duſſow*, at the mouth of the *Boryſthenes*.

Of *Muſcovia*.

Muſcovia is the vaſteſt Empire in *Europe*, 3300 miles long, and 3000 broad: The whole Countrey is over ſpread with Woods and Lakes. The People for the moſt part fat and Corpulent, ſtrong of Body, and good proportion, only Great Bellyes, and Broad Beards, are in Faſhion. The Women love not their Husbands, unleſs they beat them.

They only teach their Children to Write and Read.

They wear long Robes, under which they have cloſe coats down to their knees, but they tye their Girdles under their bellyes.

The Religion of the *Muscovite* is of the Greek Church; all their Images are in flat painting.

The Grand Duke bears the Title of *Czar*, as much as to say *Cezar*. The Habits which he is said to wear makes him look like a Priest.

The Embassadors of Forreign Princes are at the greatest trouble in the World to give him his right Titles.

One of his Pedecessors was so Barbarous, that he caused the Hat of a French Embassador to be nailed to his Head, because he refus'd to be uncovered in his presence: but Queen *Elizabeth* sent an Embassador thither soon after, a man so Couragious, that he stood also in his presence covered, and undauntedly told the *Czar* to his Face; that if he touch'd a hair of his Head, his great Mistris would make his Empire to tremble; and made the Titles of his Mistris, the Queen of *England*, Exceed the number of the *Czars* Tittles; Stiling her *The Most high and Mighty Monarch*: ELIZABETH *Queen of England, France, and Ireland, Northumberland, Westmorland Cumberland, York-shire, Lancashire, Cheshire*, &c. Runing through all the Countyes and shires in *England* the length of which Titles Amazed the *Czar*; and he acquainted the Embassador, That he had a great Esteem for her Majesty and for her Subjects; and declared his willingness to suffer her Subjects to trade in any part of his Empire, without paying any Dutyes, which great Priviledge was Continued to our English Merchants, untill the Martyrdome of King CHARLES the first, of glorious Memory; he mightely resenting that Horrid Act; and that Priviledge hath not been yet regained.

Mus-

of Europe. 55

Muscovia is Divided into two parts, the Northern and the Southern: *Mosco* is the Capital City, and the Residence of the *Czar*.

There are at this present two *Czars*, who not long since sent two Embassadors, one from each, to his Majesty of *Great Britain*: 1683.

Of *Turky* in *Europe*.

THat part of the *Ottoman* Empire which is *Turky* in *Europe*, Containes the greatest parts of *Hungary*, and all *Transylvania*, *Moldavia*, *Bessarabie*, *Walachia*, *Bulgoria*, *Servia*, *Bosnia*, *Sclavonia*, *Albania*, *Romania*, *Macedonia*, *Thessaly*, *Epirus*, *Achaia*, and *Morea*. with the *Ægean* and *Jonian* Islands, And in the year 1683 the Grand Segnior brought an Army of 100000 men in *Austria*, where with he thought to have Conquered, not only the small part of *Hungary* that is left to the Emperor, but the greatest part of the Empire also; The Conduct of which army was Committed to the Grand Visier, and past by all the strong Cityes in *Hungary*, as *Raab*, *Komorra*, &c. and sate down before *Vienna* the 8th of July. in hopes to have taken it by Surprize.

The Imperial Army under the Conduct of the Duke of *Lorrain*, their General, very oppertunely Convey'd his Infanry into the City, consisting of 15000 men; who having Count *Staremburg* for their

their Governour, a man famous for his Valour and and Conduct, that the great fury the *Turks* shewed in their several Assaults Redounded to their own loss, and the encouragement of the Besieged; and were repulsed with a very great slaughter; some say with no less then 15 or 20000 men, whose dead Bodies proved so great an Annoyance to the *Turkish* Army, that the Grand Visier desired three days Truce to bury his dead; which being denyed by the Worthy Governour, caused him to remove his Camp at a further distance, for fear of infecting his Army; and finding himself deceived in his Intelligence; (the Garrison being stronger than he was informed) he proceeded to a formal Siege, which he Continued for the space of nine Weeks; with more then ordinary Toyl and Labour, working with his Mines day and night, in hopes to have Carryed the place before any Relief could come to them; and having sprang several of them by which the place was Assaultable, he stormed it several times, but was still beaten off with great loss, and having at last sprung one under the Baston *Lobel*, he continued his Assault with greater fury and vigour, though without success; so that after so long a time, and so many Continued Assaluts, the Governour finding himself reduced to great Extremity, made the Appointed Sign to the Duke of *Lorraine* for Relief: Whereupon the King of *Poland* being joyned with the Emperiall Forces, made up an Army of 85000 Men, Horse and Foot.

It was Resolved at a General Councel of Warr, to Attaque the Turkish Camp upon the 12th of *September* new stile; And Accordingly it was put

in

in Execution. The King of *Poland* having the Attacked the Grand Vifiers Quarters, The Duke of *Lorraine* on the fide of the Baftion of the Court; and Prince *Waldeck* that which leads to the *Scotifh gate*. The Attaque being warmly begun, and Count *Staremburgh* Sallying out at the fame time, foon put the Infantry into diforder; upon with their Horfes fled, and with them the Grand Vifier, leaving them to the Mercy of the Chriftians, who cut them all to pieces, and remained Mafter of their Camp, with their Tents, the Pavilion of the Grand Vifier, Their Artillery, and Ammunition, and a Confiderable fum of Money; in this Action the Englifh that were there, behaved themfelves very Gallantly to the great Commendation of our Nation.

After the relief of *Vienna*, the two Armies under the King of *Poland*, and the Duke of *Lorrain* continued the Purfuit, and with great Succefs, and regained all the ftrong Holds in *Hungary* that were poffeffed by the *Turks*; and in all this great Action the *Turks* did not Rally their Army to Face the Chriftians: But did Continually fly before them,

THE

THE
General Description
OF
ASIA.

ASIA is a very Confiderable part of the World, in regard, Firft, That Man was therein Created; Secondly, Here our bleffed Saviour was born, wrought his Miracles, and Suffered for the Redemption of the World; Thirdly, Here was Tranfacted the moft Remarkable Occurences mentioned in the Old and New Teftament, and from hence all Nations of the World, and all Arts and Sciences had their firft beginning.

Many are the Religions here Followed; but the Jews, Mahometans and Idolaters, excel the Chriftians in number.

Mahometifme is received by the four Principal Nations of *Afia*, viz. the *Turks*, *Arabians*, *Perfians*, and *Tartars*.

The *Turks* gives moft Liberty, the *Arabians* are

are the most Superstitious, the *Persians* most Rational, and the *Tartars* most simple.

The Turks follow the Doctrine of *Omar* and have their Partriarch at *Badgat*.

The Persians follow the Doctrine of *Haly*, and have their Partriarch at *Ispahan*.

The Greeks also have their Partriarchs here, one Residing at *Antioch*, and the other at *Jerusalem*.

As to the Division of *Asia* from *Europe*, it hath already been discoursed of, in the Description of *Europe*.

The Principal Seas within the Land are the *Caspian*, the *Euxine*, and the *Persian Sea*.

The Principal Rivers are the *Euphrates*, *Tigris*, *Indus*, *Ganges*, *Crocas*, *Kiang*, &c.

The Air of *Asia* is almost every where Temperate, and abounds with Gold and Silver, Precious Stones, Spices, and Silks.

Asia is in Subjection under four mighty Monarchs viz. The *Grand Senior*, the *Sultan* of *Persia*; the *Cham* of *Tartary* (and now King of *China*) and the *Great Mogul*.

A great part of *Asia* Consists in a great multitude of Islands, of which are *Maldives*, *Ceylon*, *Sumatra*, *Java*, *Borneo*, *Aynam*, *Japan*, the *Philippines*, the *Moluccaes*, &c. And the Island of *Cyprus*, and *Rhodes*, and some others in the *Mediterranean* Sea.

Of

Of *Turky* in *Asia*.

THe *Turk* by his Puissance having over-run a great share of two of the Considerable parts of the World; Possesseth in *Asia*, these Countreys *Anatolia*, *Syria*, *Mesopotamia*, *Armenia*, and *Georgia*.

1. *Anatolia* vulgarly *Natolia*, and Anciently *Asia* the less, a Countrey once of great Fertility, but now waste and almost desolate, there were the seven Churches to whom St. *John* Directs his Book of the *Revelation*, as *Ephesus*, *Smyrna*, *Pergamus*, *Sardis*, &c.

Formerly the Air was Exceeding sound and Healthfull, now every six years the Pestilence destroys Millions of People.

The *Mahometan* Religion is chiefly professed in most places thereof.

Ephesus once famous for the Temple of *Diana*, of 425 foot in length, 220 in breadth, supported with 127 Marble Pillars, 70 foot high, 220 years in building, seven times fired.

Smirna, a place of great Plenty, the Soil abounding with Wine and Oyl.

The heats are very Excessive in Summer, (and would be unsupportable, were it not for the Breezes that come off the Sea about 10 a clock, and continues till evening,) and is followed with the Pestilence and Malignant Fevers.

of *Asia*.

Pergamus and *Sardis*, have been Royall Cities, *Pergamus* is famous for the wealth of *Attalus*, for the Invention of Parchment, and the Birth place of *Galen*, for its Tapeſtry, and for its being one of the ſeven Churches.

Sardis for the Reſidence of the Kings of *Lydia*, and alſo one of the 7 Churches.

The moſt Renowned Mountaines of the leſſer Aſia are *Taurus*, which divides *Aſia* into two parts, and is the moſt famous Mountain in the World for its heigth and length.

Euphrates divides *Armenia* and *Meſopotamia* from *Aſia Minor*, *Syria*, and *Arabia*.

Tygris, deſcends from the *Georgian* Mountaines falls into divers lakes, loſes it ſelf divers times in the Earth cutts through the Mountaines and divides *Meſopotamia* from *Aſſyria*, waſhes the Ruines of *Nineveh*, receives the branches of the *Euphrates*, and diſembogues it ſelf into the *Perſian Gulf*.

Syria, and *Phenicia*, Two Adjoining Provinces wherein is Mount *Libanus*, famous in Scripture for goodly Cedars: The Chief Cities *Damaſcus*, *Antioch*, *Tyre*, and *Sidon*, and thoſe now more renowned for Commerce are *Aleppo* and *Scandaroone*. The Couatrey abounds with Wheat, Oyl, Balm, &c.

Canaan or *Paleſtine*, in length but 200 miles, and in breadth not above 50; yet ſo Fruitful that we read in Holy writ, it once ſuſtained 1300000 fighting Men, beſides the tribes of *Levi* and *Benjamin*, but now it is nothing ſo plentiful; 'Tis now under the *Turk* Inhabited by mixt People, pretending to Christianity, shewing divers of the places mentioned in the Scripture.

Mesopotamia or *Diarbeck* lying between *Euphrates* and *Tygris*, the Soil is exceeding Fertile, and from hence supposed to be the place where the Garden of *Eden* was.

Upon *Tygris* stands *Babylon*, now called *Bagdat*, where happened the Confusion of Languages after the Flood, and is only a great Ruinous heap near which place stood the Tower of *Babel*.

Babylon was built by *Nimrod* much Augmented by *Nebuchadnezzar*, it was taken by *Cyrus*, *Darius*, and by *Alexander the great*, who died there.

In the year 1638 when *Amurath* the fourth retook it from the Persians, he caused three Men out of every Tent throughout his Army, to be cast into the Mote, and over them a vast number of Bavins and Woolsacks, that he might the more easily assault the place.

Armenia now called *Turcomania* taking its name from the *Turks*, who being a People of *Schythia*, and too populous to get food sufficient in so Barren a Country broke through the *Caspian* Sea, and seated themselves here in the year *Anno Domini* 844.

Georgia, not so called from St. *George* whom yet here they greatly reverence but from the *Georgi* who long ago Inhabited the Adjacent Countreys, the People now though Tributaryes to the Turk, Profess Christianity, and agree in most points with the Greek Church: It is seated between the *Euxine* and the *Caspian* Sea.

Of

Of *Arabia*.

Rabia is bounded on the East with the *Perfian Gulph*, and *Chaldea*, and on the South the *Ocean*, and on the West the Red Sea, and some part of *Egypt*, and on the North the River *Euphrates*, with some part of *Palestine*.

It is Commonly divided into three parts, viz. *Arabia* the Stony, *Arabia* the Desert, and *Arabia* the Happy.

Arabia the *Stony* lies near the Holy-Land.

Arabia the *Defart*, lies near *Chaldea* and the *Euphrates*; *Arabia* the *Happy* advances it self between the Red Sea and the *Perfian Gulph*, which divides it from *Perfia*; and this part is the greatest, and Richest, and best Inhabited of all.

Arabia the *Stony* hath for its chief City *Herat*, which signifies a Rock, whereon it was built, with an advantagious Scituation, a place of great strength.

On the Coast of the Red Sea is the Castle *Tar*, a Walled Town and a Port, very famous, and it is believed the Israelites having passed the Red Sea, Entered the Desarts this way; and it is likewise a Passage, where the Carravans stop at their return from *Mecca*.

Arabia the *Defart*, so called by reason of the vast Sandy Desarts, and the uninhabitablenefs thereof, of, scarce offerding either food for Man or Beast: So that those which travel those Countreys are forced to carry their Provisions with them, and guide themselves to the place designed by the help of the Stars, or Marriners Compass, as they do at Sea, and go in great Companies for fear of being robbed or rifled by the wild Arabs.

The People are much addicted to Theft by which they get their living, being stout and warlike Men; their Chief food being Venison, Milk, and Herbs; they go half naked; their Wives they hire for what time they please,

Arabia the Happy may properly be so called by Reason of the Fruitfulness and Richness of the Soil, which Produceth plenty of Corn, Wine, Fruit, and Odoriferous Spices; great increase of Cattel; also abounding in Gold, Pearles, Balsom Myrrhe, Frankinsence, and several sorts of Drugges

These People are very Faithful and Punctual in their Promises; here are great quantities of Ostriches which for the most parts abide in the Desarts.

In the Province of *Hagiaz*, the Chief Cities towards the Red Sea are *Medina*, and *Mecca*, this last being the Birth-Place of *Mahomet*, and the other his Burying-Place.

Medina, though Scituated in a Barren and Desolate Place, adjoyning on *Arabia* the *Stony*, yet by Reason of, its being the Sepulcher of *Mahomet*, is become a fair City, containing about 6000 Houses, being a place of great Trading and resort, by Reason of the Pilgrims, which flock thither to pay their Blind Devotions.

This Sepulcher or Tomb wherein their Prophet lieth;

lieth, is Enclofed within an Iron grate, and Covered with green Velvet, having the fupply of a new one every year from the *Grand Seignior*, and the old one becomes the fees of the Priefts, which they fell in fmall pieces and fhreds for great Reliques to the Pilgrims, which brings them in great Revenues.

In this Temple are about 3000 Lamps of Gold and Silver, wherein is Balfam and fuch other rich Odours, Oyntments, and Oyls, which are Continually kept burning.

The People for the moft part are of a mean Stature, Lean Swarthy Complexioned, Effeminate voices, very Swift of foot, and very Expert in the Bow and Dart.

The Horfes are alfo little and lean, and fpare in feeding; yet Couragious, Swift, and of great Labour.

The People are almoft all *Mahometans*, except fome Greek Chriftians, towards the Mounts of *Sinai* and *Horeb*; likewife towards the *Red Sea*, and in the Defarts of *Arabia the Stony*, and *Arabia the Defart*; *Arabia the Happy*, is the unhappieft, by having the feweft.

Of *Perfia*.

THe Kingdom or Empire of the *Sophy* of *Perfia*, is one of the greateft and moft famous of all *Afia*; it Extends from *Tygris* and *Euphrates* on the weft west, almost to the River *Indus* in the East; And from the Gulph of *Persia*, and the *Arabian* and *Indian* Sea, which bounds it on the South, unto the River *Gehon* and to the *Caspian Sea*, which are its Bounds of the North; so Containing about 600 leagues in length, and 500 in breadth.

The Persian Monarchy began under *Ninus*, and lasted under 30 and odd Kings 13 or 1400 years, ending in *Sardanapalus*.

It divided it self into *Medes* and *Babylonians*, afterwards the *Persians* made themselves Masters of it.

Alexander the Great held this Empire but few Years; and dying, it was divided amongst many of his Captaines, who at last took upon them the Title of Kings.

Hispahan the *Metropolitan* City of the *Persian* Monarchy, seated in the *Parthian Territory*, which in its Situation is pleasant and delightful, and in its Soil Fruitful, its air Serene and Healthful, and for bigness is now become the greatest City in all *Persia*; whose Walls are in Circumference a reasonable Days Journey, containing not less then 75000 Houses.

The Inhabitants do all their affairs on Horse-back, as well Publique as Private, in the buying and vending their Comodities: But the Slaves never Ride, which makes the difference between them. The Native Comodities of *Persia* are Gold, and Silver, Raw-Silk in such great Quantities that they furnish most part of the East.

The heats in these parts in the Summer season are so great, especially towards the South part of the Mountains, that the Inhabitants are forced to forsake the

the Cities, and retire into the Mountains for coolness.

The *Persians* are of low Stature, yet have great limbs and strong, they are of an Olive Colour, hawk'd Nose'd, and Black Hair'd, which they shave every Eight Days; they paint their Hands and Nails of a Reddish Colour.

In their Habits they follow much of the *Turks*; their Clothes have no proportion to their Bodies, hanging loose and large, much in the Fashion of the Women; their Garments they Gird about their Waists with a Scarf; Their Turbants are Red. The Women in their streets, go with white Vailes over their Faces, down to their knees; they are very Neat and Rich in their Clothes and Houses.

The *Persians* are very Strict, Superstitious, and Ceremonious in their Religion, as the *Turk* is. They Bury their Dead three hours after they are departed, Except they Dye in the night: They wash or bathe the Bodies of their Dead before they are interred, in a great Cistern, which they have for the same purpose near the *Mosque*, to which place they are Carried on a Bier in their Clothes, and after they are stript and washt, they put them in clean Linnen, Anoint them, and so bear them to the Grave, being accompanied with his Friends, Relations, Servants, &c.

The King of *Persia* Governs by an absolute Power, disposing of the Lives and Estates of his Subjects, as best pleaseth him, making his Will his Law, not daring to Murmur, though his Actions are never so unjust.

The Administration of Justice is decided by the

King, but firſt tryed by the Secular Judges, who Examine the ſame, and deliver up their Opinion to the King.

They have ſeveral Severe and ſtrict Puniſhments, which they inflict upon the Offenders, according to the Hainouſneſs of the Crimes; for ſome Offences they cut off their Ears and Noſe, ſometimes their Foot and Hands, for others to be Beheaded; for ſome again, they are tyed between two boards and ſo ſawed aſunder, with ſeveral other cruell Deaths, too tedious to name.

In their Military affairs they are well Experienced, their Army conſiſting only of Horſe, who have for their Armour Darts and Javelins, yet have they ſome in the Nature of our Dragoones.

They have great quantity of all Sorts of Cattle, Grain and Fruits; amongſt their Fruit Trees, they have great quantities of black and white Mulberry-Trees, which grow not above five or ſix Foot high, which ſerves for Food for their Silk Worms, which makes them great ſtore of Silk.

Of the Empire of the Great Mogul.

THe Great Mogul, is one of the greateſt and moſt Powerful Princes of *Aſia*, whoſe vaſt Empire Comprehends the Greateſt part of the Continent of *India*. In which large Territories there are

are several *Indian* Kingdomes Tributary to him; He is esteemed the Richest Prince of the World; *Sha Johan* who Raigned 40 Years, left him, behind him, five Millions of Livers: And the Throne that he made cost one hundred and Sixty Millions, and Five-Hundred-Thousand Livers, besides six other Thrones, set all over with Diamonds, Rubyes, Emeralds, and Pearles.

He is able to bring into the Field One-Hundred-Thousand Horse, and Two-Hundred-Thousand Foot, and two or three thousand Elephants.

The Great *Moguls* Ordinary Guard Consists of about twelve-Thousand Men, besides Six-Hundred of his Life Guard; he never stirs abroad to Hunt, take the Air, or the like, without the Atendance of Ten-Thousand Men of all Degrees; besides, to make his State the Greater, there are one Hundred Elephants, Richly trapt and covered with Scarlet Velvet, or the like; on each of these Elephants there are seated two Men, the one to guide him, and the other Supports a large Banner of Silk richly Embroidered with Gold and Silver; after these Hundred Elephants comes the *Mogul*, either mounted on an Excellent Horse, or else in a Coach or Sedan, attended by his Nobles and other Courtiers: After whom comes about Five-Hundred Elephants, Camels and Waggons, which are to Carry the Baggage; For he Commonly Encamps in the Field, to have the Benefit of the Coolness of the Air: The *Mogul* Celebrates with great Pomp and State the first Day of the Year.

The Emperour himself is a *Turkish Mahometan*, although the most part of his Vassals are *Pagans*;

For as there are several sorts of People, so there are divers sorts of Religions amongst them. The Country is very Fertile, yielding great store of Corn and Rice; and all Sorts of Provisions are very Cheap, and the Inhabitants very Sober and Temperate.

It is conceived to be the largest Country of any one name in the World, (Except *China* and *Tartaria*.)

The whole Countrey is Divided into two *Peninsula's*, one *Peninsula* is accounted on this side the River *Ganges*, called *India Intra-Gangem*, and the other *Peninsula* beyond the *Ganges*, called *India-Extra-Gangem*, of each we shall take a Brief Survey.

Of *India* on this side the *Ganges*, called *India-Intra-Gangem*.

INdia-Intra-Gangem, is bounded on the East with the River *Ganges*, till the fall thereof into the Sea; And after, that with that large and spacious Bay, called the Gulf of *Bengala*. On the West, with the Territories of the *Persian Empire*, and part of the *Arabian* Seas; On the North where it is broadest with Mount *Taurus*; On the South with the *Indian Ocean*, ending in a sharp point or Promontory, as you may see in the Map.

We Divide this *Peninsula* into ten parts or Divisions, being under several Kings, Governments, and Religions. In the Kingdom of *Cuncan*, are *Goa*, *Visa-*

Visapore, *Dabul* and *Rejapore*, *Carapatan* and *Mongrelia*: *Goa* is a City, as fair, Rich, and of as great Traffick as any in the East-*Indies*, being Situated on an Island of the same name, taken by the *Portugals* in the Year 1510, and have since that Established themselves so Powerfully there, that their *Vice-Roy*, Bishop, and their Council, for the East-*Indies*, have their Residence there; Their streets are large, their Houses fair, especially their Palaces and Publick Buildings, which are very Magnificent, their Churches are Stately, and Richly Adorned; The City is 15 Miles in Compass. The *Portugals* live here in all manner of Delight and Pleasure.

Here on this Coast is also Situated *Surrat* and *Bombay*, two great Factoryes of the *English* Merchants.

The whole Countrey is under the Government of the *Suvagee*, who is a Tributary to the great *Mogul*. The People bloudy and treacherous, addicted much to the Art of Poysonning, and do Fight commonly with Poysoned Arrows.

A notable Instance of their Treachery I shall give you; Which was in the Year 1683. There was three of their Ships and four Grabs, man'd with 1500 of their Men, which the *Suvagee* afterwards (by way of Excuse) pretended to be *Arabians*, and took the Ship to be a *Portuguese*, and as a token of their Resentment Imprisoned the Chief of them; They set upon one of our East-*India* Ships, call'd the *President*, Captain *Jonathan Hide* Commander, whom they Attaqued very desperately, and was as Briskly received by the *English*, and as bravely Repulsed with a great Slaughter of the *Indians*: And

E 4 Re-

Recovered their Ship, to the great Comendations of all that were there in. Which brave Souls had not the Happiness to Enjoy the Honour of that Action in their own Countrey. The Ship being unfortunately cast away in *February* following, and all their Men lost but two, as they were coming into the *Chanel*, to the great grief and sorrow of the Merchants and Owners, as well as their Particular Friends and Relations.

Malabar, extends its self from the River *Gangeraco*, to *Cape Comarine*, 300 miles in length, but is not above 50 in breadth, and ending towards the *Cape* in a Point; The Countrey is more Populous for the bigness then any in *India*, enjoying a very Temperate Air, and a Fruitful Soil, well Watered, and indented with many Creeks; The Ground unfit for Corn but Plentiful in Rice, and all manner of Spices, as Ginger, Cinnamon, Cassia, Pepper, and most Excellent Fruits.

The People on this part of *India*, are of a blacker Complexion then the rest of the *Indians*, well Limbed, wearing their Hair long and Curled: About their Heads they wear a Hankerchief, wrought with Gold and Silver, and about their middle a Cloth, which hangeth down to conceal their Nakedness.

The Natives on the whole Coast of *India*, are of Nature Treacherous and Bloudy, more properly to be termed Desperate, than Stout and Valiant; in their Wars they use Poysoned Arrows, as they do in their Pyrating and Thieving, both by Sea and Land.

The Kingdom of *Bisnagar*, seated in the *Bay* of *Bengala*, between twelve and fifteen Degrees of North Latitude, in which Kingdom is the great Factory

Factory of our East-*India* Company, called *Fort St. George*; almost all the People are Idolaters, some *Mahumetans* and a few *Catholiques*. Amongst the Customes of those Heathens, they have the Inhumane Custome for the Women, to Burn themselves with the Corps of their deceased Husbands, (in this manner) the Husband being Dead, the Wife prepares her self for her Funeral Habite in Transparent Lawn; Her Nose, Ears, and Fingers she Adorns with Precious Stones, in one hand She holds a Ball, and in the other a *Nosegay* of Flowers, both as Emblems of *Paradise*: And being thus Habited: She is Accompanied to the place by her Relations, Friends, and Acquaintance, and all the way Going, the Priest denotes the Joys She is to Possess, together with the Assurance of Enjoying her Husband, which does Excite her to *Valour*; so that when She cometh to the Place, seeth the Flame, and the Carcass of her Dead Husband, whom She longeth to be with, and being as it were Transported with Joy, She takes leave of her Friends and Relations, and Jumps into the Flame, in which the Corps of her Husband was first put, both which is soon Reduced to Ashes, during which time they have several sorts of *Musick*, to Drown the Cryes of the poor Wretch, casting in store of sweet Wood and Oil into the Fire, to take away the Unsavory smell.

Of the Island *Ceylon*.

THe Island *Ceylon*, lyeth to the Eastward of *Cape Comarine*, of an Oval form, divided from the *Main*, by a shallow *Chanel*. The Havens Refreshing it with their Dews; The Air with Fragrant freshness.

The Land stored with whole Woods of *Cinamon*, besides Fruits, Lemons and Oranges, &c. Of Fowls and Beasts great Plenty, both Wild and Tame; It so abounding with all Contentments, that the *Indians* hold it to be a Paradise; In this *Island* is a mighty high Hill, called *Adam's Pico*, reputed to be seven Leagues high.

The People for the most part Tall and Strait of Body, in matters of Religion they are generally Idolaters, very Active and well Skilled in Jugling. This *Island* is Divided into several Kingdoms, as of *Candie*, &c. And a City of that Name, the Seat of the King. This *Island* is memorable for the 20 Years Captivity of *Captain Knox*, and of his Miraculous Escape, which when he Returned to *England*, he wrote a large Treatise of the Nature of this *Island*, and the Manners and Customes of the Inhabitants, (to which) I refer you.

Of the *Maldive Islands*.

NEar unto the Coast of *Malabar*, lieth a great Number of Islands, called *Maldive*, from *Mal*, the City of those *Islands*, and *Diva* which signifies an *Island*; They are Divided into 13 divisions, divided by certain *Channels*. From hence the King of *Maldives* terms himself King of 13 Provinces

vinces, in which are Contained Eleven-Thousand Islands, as is Reported by some that have been there; here they take Multitudes of little shells, called *Couries*, which pass in many places for Money. The King Resides in the *Isle* of *Mal*, which is one of the greatest; but not above a League and a half In Circumference, but it is a most Fruitful place, strangers frequent it, because of the Court; There Grows on these Islands neither Rice nor Wheat, yet all sorts of Provisions are Cheaper here then in other parts of *India*; There are here abundance of *Cocoe* Nuts; they have little Beef or Mutton, no Dogs (for they Abhor them,) they have great store of Fish. The Inhabitants are very Industrious and Sober People.

Of *India* beyond the *Ganges*, called *India-Extra Gangem*.

Ndia Extra-Gangem, is that part of the great Continent of *India*, which lieth on the East-side of the River *Ganges*; In this *Peninsula* are a great Number of Kingdomes, but I shall only discourse of the Principal ones that are Contained in the Table, which are *Arracan, Pegu, Martaban, Siam, Cambodia, Couchin, China, Malacca*, and *Tanquin*.

The Kingdom of *Arracan* is Situated on the East-side of the *Bay of Bengala*, extending it self from the *Tropick of Cancer* to the sixteenth Degree of North

North Latitude, it being a brave and Fruitful Country.

The Kingdom of *Pegu*, when in its Splendor, was so Rich and Powerful, that some would equal it to *China*. They have in many places Mines of Gold, Silver and Precious Stones; Besides Grains, Fruits, Herbs, Fowl, and Beasts, which are here found Excellent. And without doubt it is the Richest Country of all the *Indies*, and for the same Reason one of the best Peopled and most Powerful.

The Chief City of this Kingdom is *Pegu, the Metropolis*; The Houses well Built; The *Palace-Royal* is Seated in the midst of the City, having its partitular *Wall-Moat* and other Fortifications.

The Habit of the Natives is but mean, contenting themselves for the most part with a piece of Linnen, to Cover their Nakedness; They all black their Teeth, because they say Dogs Teeth are white; They are Generally all *Pagans*; Those that Marry, Buy their Wives of their Parents.

The Kingdom of *Martaban* towards the Gulf of *Bengala*, is Contiguous to *Pegu*, to which it hath been Subject, at present to *Siam*; this Kingdom hath many Ports frequented for Trade, for besides its Grains, Fruits, Oyls, and Medicinal Herbs, it is Rich in Mines of Gold, Silver, Copper, &c.

The Kingdom of *Siam* hath several Cities of Note, but we shall (for Brevity sake) only treat of *Siam*, as the *Metropolis*, being a City of large Extent, and of so great strength, that in the Year 1567. they stoutly defended themselves against an Army of 1400000 Fighting Men, which the King of *Pegu* brought against them, for twenty Months together. The Comodities of this City and Kingdom are Cotton,

ton, Linnen of several sorts, *Benjamin Lacque*, of which they make Excellent hard Wax.

The Kingdom of *Cambodia*, which lieth between the Gulfs of *Siam* and *Couchin-China*. The Principal Cities *Ravecca* and *Camboja* ; The People in their manners and Customes, Resemble those of *Siam*, whose Subjects they have been.

The Kingdom of *Couchin-China*, lieth to the Eastward of *Camboja*, its Name signifies west *China*, and was supposed formerly to belong to *China*, whose Language, Manners, Customes, Government, Religion, and other Ceremonies, they yet retain. All the Countrey is Fertile, abounding in Rice, Fruits, and Herbs. They have Gold, Silver, Silk, Porcelain, and many other valuable Commodities; The Air Healthful, and the Soil so Abundant in all things, that the Inhabitants know nothing of Contagion, or Famine. They are Courteous, Liberal, kind to Strangers, and Faithful in their dealings.

In the *Peninsula* of *Malacca*, are diverse Kingdoms, which are Tributary to *Siam*, Except the City of *Mallacca*.

The whole Country is well Traded, by Reason of its *Archepelago*, which contains several *Islands*, and of its *Isthmus*, which facilitates the Transportation of Merchandise, from one Sea to another; and of its Ports, which are Commodious.

Patane, within few Years is grown Famous; The Kingdome being frequented by diverse Nations, perticularly the *Chinois*, who bring thither Parcelain, and divers Manufactures and Instruments of Husbandry; The Soil is good, producing Fruit every Month in the Year; Their Hens, Ducks, and Geese, often lay Eggs twice a Day, *Ibor*

Ihor is Situated on the utmost point of the *Peninsula*, its Chief City was taken and ruined by the *Portugals* in 1603. who took from thence Fifteen-Hundred brass Cannons.

The Kingdom of *Tanquin*, divides *China* from *Couchin-China*, and hath about 150 Leagues of Sea Coast. This Kingdom contains 7 Provinces. The King of *Tanquin*, Ordinarily entertaines a Militia of 50000 Men.

The Land hath Beautiful Plaines, and Watered with many great Rivers; They have neither Asses nor Sheep, but many Horses, Elephants and *Rhinocerotes*, whose Flesh, Skins, Bones, Teeth, Nails and Horns, serve for *Antidotes* against Poyson; The Catholique Religion was so Introduced some Years past, that there was estimated to be more then 200000 Christned Souls; Two-Hundred great Churches, and great Quantities of Chapels and Oratories, but since there hath hapened great Changes; in those Kingdomes the *Portugals* have several Towns and Cities, by which they have a great Revenue.

Of the *Island* of *Sumatra*.

THe *Island* of *Sumatra*, lieth on the North of *Java Major*, and a long the West-side of *Mallacca*, the length thereof Extends from the North-West to the South-East, about 800 miles, and is 200 in breadth.

The Air is very hot and unwholsome, by Reason of

of the grofs Vapours, drawn from the many Fens and Rivers, which are found to be in it. The Soil not Capable of fuch Grain as in other places, except Rice and Millet; it Affords great plenty of Wax and Honey, ftore of Silks and Cottons, Rich Mines, not only of Tin, Iron, Copper, Sulphur, and other Minerals: But of Gold fuch quantity, that fome conceive this Ifland to be *Solomons Ophir*, for one of the Kings thereof wrote a Letter to King *James*, wherein he hinted the Riches of his Country Abounding in Gold, and that he had in his *Dominions* a Mountain of Gold, King of which Golden Mountain, he Intituled himfelf in his Regal Stile, his Title is alfo King of *Achem*. His Government is Abfolute, and meerly Arbitrary, executing what he hath a mind to, without form of Law. So Cautelous, that without his *Placard*, no Stranger can have Ingrefs into his *Dominions*, nor is Admittance to his Prefence granted to any whom he firft fends not for, by an Officer with a Gilded Staff; He is conceived to be ftrong, both by Sea and Land; his Country Populous, his Elephants many and well Trained.

Of the Ifland Borneo.

THe Ifland *Borneo*, is Situate under the Æquator, and is Adjudged to be more then 2200 miles in Circumference. The People Generally more white then the reft of the *Indians*, of good Wits and Aproved Integrity.

The

The Country is said to be provided Naturally, with all things Neceffary; But unfurnished with *Affes*, *Oxen*, *Herds of Cattel*, except only Horfes, and thofe but fmall of Stature; the great Riches of it, being *Camphir*, *Agarick*, and fome Mines of *Adamants*.

The Ifland is divided between two Kings of two Religions; The King of *Borneo* and his Subjects are all *Mahumetans*, and thofe of *Laus* ftill Remaining in their Ancient Gentilifme. Thefe think the Sun and Moon to be Man and Wife, and the Stars their Children, Afcribing to each of them Divine Honours, to the *Sun* efpecially, whom they Salute at his firft Rifing, w'th great Reverence.

Their Publique bufineffes are Treated of commonly in the Night.

The King of *Borneo* keeps the Greater State, not to be fpoken with, but by the Mouth of fome of his own *Interpreters*, and in his Palace Served by no other Atendants, than Maids or Women.

Of the Ifland *Lucon*.

THe Ifland of *Lucon* begineth at the thirteenth Degree, and continueth up to the Ninteenth Degree of North Latitude, lying South from *China*. *Manilla* is its Chief City, well Built after the Modern way, and its Houfes are of free Stone, Built by the *Spaniards*; this being one of the *Philippine* Iflands; So called from *Philip* King of *Spain*.

The

The Governour, or Vice Roy of these Islands, as also an Arch Bishop, who hath a Spiritual Jurisdiction over all these Islands which he exercises by three Suffagan Bishops.

This City is very populous, here commonly residing 15000 Chinois, besides *Iponeses* and *Spaniards*, which drive a Trade in several good Commodities.

Of the Island Paragoa.

THe Island *Paragoa* stretcheth it self South-west, and North-East, in length above 100 Leagues, not having above Ten, Fifteen, or Twenty Five in breadth, it begins almost at the Eighth Degree, and ends not till the Eleventh Degree of North Latitude. It is said, that it bears Figs as thick as ones Arm. Its King is a Vassal to him of *Borneo*.

Of China.

CHina is bounded on the North, with A'tay and the Eastern Tartars, from which Continued with a Chain of Hills, (part of those of *Ararat*) and where that Chain is broken off, or interrupted with a great wall, extended 400 Leagues in Length.

On the south part with *Couchin*, *China*, and partly with the Ocean; on the East with the Oriental Ocean, and on the West with part of *India*. It is said to contain in Circuit 3000 Leagues, Containing not less then Sixteen Provinces at this day. The Country is very Rich and Fertile, so that in some places they have two and in some three Harvests in a year.

The People are for the most part of a Swarthy Complexion, short nosed, black eyed, and very thin beards; they wear their Garments very long, with long loose sleeves, and their hair long; they drink their drink hot, and eat their meat with two sticks of Ivory or Ebony.

The Son is bound to follow his fathers occupation; The People are good Artificers, Ingenious and Excellent in all things they take in hand, as the Purcelian Dishes, curious Carvings, and the fine painted works which comes over from thence doth manifest. They are effeminated with ease and pleasure; and are not much given to Wars.

Of both Sexes, there is thought to be Contained in this Country not fewer then Seventy Millions. In Matters of Religion, are generally all Gentiles.

The Forces which this King is able to draw into the Field, must needs be infinite, Considering that incredible number of Subjects under his Command, for whereas *France* is thought to Contain Fifteen Millions of People, *Italy* with the Isles, as many, *Germany* with the *Switzers* and *Belgick* Provinces about that Proportion; *Spain* not above Seven Millions, and the Kingdom of *England* and *Wales* not above Five Millions which

is

Of Asia.

is in all 57 Millions, his people is 13 Millions more then all these put-together. The Government of this Kingdom is Tyrannical, there being no Lord but the King, no Title of Dignity or Nobility known amongst them, nor Toll or Duty paid to any but to him.

The Dignity of the Crown is Hereditary, falling to the Eldest Son after his decease. The King, they highly reverence, calling him the Son of Heaven, and the Son of God, &c.

The great City of *Pequin*, now the Seat of the King is of a vast bigness, Containing within its Walls 3300 *Pogodes* or Temples, wherein are continually sacrificed a great number of Wild Beasts and Birds. In the Walls which encompasseth this City are 360 Gates, to each of which is joyned a small Fort, where a Guard is continually kept as also a Register to take the names of persons, that pass thereat, each of the streets having its Captain and other Officers who are to look after the same, and every night to shut up the Gates; Here are about 120 Aquaducts and Canals upon which are near 1800 Fair Bridges sustained on Arches. This City is the Residence of the King, when he is in the Northern Provinces as *Nanquin* is in the Southern.

Of Tartary.

TArtary is seated on the most Northern part of *Asia*, and extends it self from East to West; from the River *Volga*, and *Oby* (that separates it from *Europe*) unto the streight of *Jesso* which separates it from *America*. Their Neighbours are the *Muscovites* on the West, on the North the *Tartarian* Sea, and the *Persians*, the *Mogolls* and the *Chinois* on the South, on the East the Straits of *Jesso* or *Anian*, not yet certainly known.

The People are of an indifferent Stature, ugly Countenances, thick Lips, Hollow Ey'd, Flat Noses, broad faced, very strong, stout, Valiant, and good Warriers, very Active, Vigilant, exceeding quick of foot; patient in all afflictions, they are very Rude, Barbarous, and revengeful, do eat their Enemies, and drink their blood, as Wine at Feasts.

Their Habit is mean, made of Course stuff, reaches but to their Knees, yet they are very Proud, and think their *Cham* to be the greatest Prince in the World.

In Matters of Religion they are generally *Pagans* and *Mahometans* The *Pagan* being the best Gentleman, being of the Elder House.

Their Food is mean and sluttishly drest; they eat Horses, and drink Mares Milk.

The Government is Tyrannical, their Great *Cham* being Lord of all, in whose Breast lyeth their Laws. Every man hath the liberty of having two or three Wives, which they never choose, but out of their own Tribe.

The Country is very Fertile in most places, abounding in Wheat, Rice, Wool, Hemp, Silk, Musk, Rubarb, great Herds of Camells and Horses, which they vend to the *Chinois*, *Mogolls*, and other Indians that comes thither to Traffick. A

A General Description.
OF
AFRICA.

CHAP. I.

AFRICA is a Peninsula so great, that it makes the third, and most southerly part of our Continent. It is bounded on all sides by the Sea, it is by the Latines called *Africa*, and the Greeks *Lybia*. It approaches so near to *Spain* that onely the Straights of *Gibralter*, divides them, and is joyned to *Asia* only by a small *Isthmus* of Thirty or Forty Leagues between the Red Sea, and the Mediterranean.

It is every where Inhabited, (though not so well as *Europe* and *Asia*,) partly by reason of the unsupportable heats, and partly for want of water in many dry Countryes.

A general Discription

As to its divisions. In the higher part of *Africa* is *Barbary*, *Billedulgerid* and *Egypt*; further South is the desarts of *Zaara*, the Country of Negroes and *Guinea*; in the higher *Ethiopia*, or under *Egypt*, are *Nubia*, *Abiffina*, and *Zanguebar*; In the lower or Inferiour *Ethiopia*, *Congo*, *Mono*, *Motapa*, and the *Cafrees*.

Barbary extends it self along the Mediterranean-Sea from the Ocean to *Egypt*, and is bounded on the South by Mount *Atlas*.

Billedulgerid lies along this Mountain, likewise from the Ocean unto *Egypt*, bounded by *Zaara*, or desart. *Egypt* is only one valley from the *Cataracts* of *Nile* unto the Mediterranean-Sea. Likewise *Zaara* the desart, the Country of the *Negroes* and *Guinea*, stretch themselves from the Ocean unto the high and low *Ethiopia*; We have divided *Ethiopia* into the higher and lower placing in the Higher, *Nubia*, *Abiffina* and *Zanguebar*; in the lower *Congo*, *Monomotapa* and *Cafrees*.

The Mountains in *Africa* are in great number and are very remarkable for their heighth, and the Mettals wherewith they abound. The most famous are Mount *Atlas*, those of the Moon, and *Siere Lione*; *Atlas* was the most famous amongst the Ancients, who believed it bounded the world on the South; The Mountains of the Moon are higher than any in *Europe*, and are alwayes covered with Snow and Ice. The largest and most famous Rivers of *Africa* are the *Nile* and the *Niger*.

The Emperors, Kings and Princes which at
present

present possess *Africa* are in very great number; The most powerful and Confiderable, are the great Turk or Sultan of the *Ottomans*, who hold all *Egyp*, a great part of *Tarbary*, and almoſt all the Coaſt that touches the Red Sea. The *Negus* of the *Abiſſines*, who poſſeſſes the faireſt and greateſt part of the Higher *Ethiopia*, the *Xeriffs* of *Fez* and *Morocco*, which have held thoſe two Kingdoms in *Barbary*, in which Country was ſituated the City and Garriſon of *Tangier*, belonging to his Imperial Majeſty of Great *Britain*, which was demoliſhed in the year 1684. managed by the prudent Conduct of the Right Honourable *George Lord Dartmouth* in the Ship of *Grafton*, and ſeveral others under his Command.

The Religions in *Africa* may be reduced to four, viz. *Mahometiſm, Chriſtianity, Paganiſm* and *Judaiſm, Mahometiſm,* poſſeſſes *Barbary, Biledulgerid, Egypt, Zaara* the deſart part of the *Negroes* and a good part of *Zangubar.* *Paganiſm* holds part of the *Negroes, Nubia* and *Guinea*, and all the Lower *Ethiopia*, with the *Cafrees,* and ſome mixture otherwhere. Chriſtianity holds in *Africa* almoſt all the whole Empire of the *Abiſſines*, and part of *Egypt,* and by the *Portugals* in their ſeveral Factories and Colonies that they have upon the Coaſts of *Africa.*

As for *Judaiſm* it is ſcattered in many Cities on the Coaſt of *Barbary,* as at *Morocco, Fez, Algier, &c.* Likewiſe in *Egypt,* and on the Confines of the *Abaſſines.*

CHAP.

CHAP. II.

Of the Country of Africa in particular.

Of Barbary.

THe People in *Barbary* are of a duskish or blackish Complexion, of Stature Tall and well proportioned, they are of an Active Disposition for Horsemanship, otherwise exceffive idle; they are very fubtil, close, perfidious, inconftant, Proud, much addicted to Luxury; and therefore by confequence very jealous of their Wives, whom they keep with great severity, and that the more according to their handsomnefs.

Their Religon is *Mahometifm*, and are for the moft part inclined to *Literature* and Arts.

The Moores of *Fez* and *Morocco* are well difposed, strong, active, and yet melancholly, they may marry four Wives, and as many Concubines as they can keep.

Here the women at the death of their friends, affemble themfelves together, habit themfelves in Sackcloath and Afhes, and fing a Funeral Dirge to the praife of the deceafed, and at the end of every Verfe howl and cry, and this they do for feven days together. *Of*

Of Billedulgerid.

Billedulgerid, or Land of Dates, hath *Barbary* on the North, from whence it is separated by Mount *Atlas*, on the south *Zaara*, on the west the great Ocean Sea, and on the East *Egypt*.

The Air is healthful, they live long, are deformed, and are held base people, ignorant of all things, are addicted to Theft, Murther, and are very deceitful, they feed grosly, and are great hunters, they acknowledge *Mahomet*.

Of Zaara.

Zaara is an *Arabian* name, and signifie Desait; The Country is generally hot and dry, it hath but little water, except some few wells, and those salt, if there falls great rains, the Land is much better; but besides the leanness of the soil, there is sometimes such vast quantities of Grashoppers, that they eat and ruine all that the earth produceth.

It is so barren and ill Inhabited, that a man

may travel a week together without seeing a tree, or scarce any grass or water.

The People are *Bereberes* and *Africans*, almost all follow *Mahometism*.

This great desart is divided into five principal parts, as is shewed in the Table and Map.

Of the Land of Negroes.

THe Negroes are People about the River *Niger*, which hath taken its name from these people, and these people from their Colour.

In this Division are placed several Kingdoms as you will find marshall'd in the Table; and what is remarkable in them, we shall briefly touch.

Some of the Kingdoms are rich in Grain, Cotton, Cattle and Gold, the Country of the *Negroes* is esteemed as fertile, as those watred with the *Nilo*; it bears twice a year, and each time sufficient to furnish them with Corn for five whole years; which makes them not sow the Lands, but when they judge they shall have need; they keep their Corn in Pits and Ditches under ground, which they call *Matamores*.

The People are generally idle and ignorant but bear great respect to their Kings.

of Guinea.

GUinea is that part of the Coast of *Africa*, which is found between the River *Niger* and the Equinoctial Line. This Coast from East to West is 7 or 800 Leagues long, and not above 100 or 150 in breadth.

The Soil of *Guinea* very fertile, and for the moſt part bears twice a year, becauſe they have two Summers and two winters. The Comodities of the Country are Gold and Elephants Teeth in great abundance, in Wax, Hides, Cotton and Ambergreeſe, and for theſe Commodities, they barter for courſe Cloath, both Linnen and Woolen, Red Caps, Freez Mantles and Gowns, and leather bags, Guns, Swords, Copper Bars, and Iron Knives, Hammers, Axes, &c.

The Inhabitants go naked, ſave about their Waſte they tye a piece of Linnen, yet very proud and ſtately, in matters of Religion, great Idolaters worſhiping Beaſts; on this Coaſt are ſeveral Factories belonging to the Royal *African* Company.

Of Congo.

TO the ſouthward of the Equinoctial Line, and unto *Cape Negroe* lies the Kingdom of *Congo*; and is ſaid to be the faireſt of the lower *Ethiopia*. The Inhabitants are naturally very ſweet, and are able and ſtrong of body, but dull and idle, their money is of grey ſhells, their Grains, Fruits, Waters, Fowl, Sea and River Fiſh are excellent, they have ſtore of Elephants, Mines of Silver, Iron, Chryſtal, Marble, Jaſpar, Porphyre, &c. They know no Hiſtory but by the Reigns of their Kings.

The moſt famous Rivers of this Kingdom are the *Zaer*, the *Lelunda*: the *Zair* deſcends from the Lake of *Zair*, from whence alſo deſcends the
Nile

Nile, The *Zaer* hath 400 Leagues Courſe, and is very Rapid by reaſon of the Cataracts or great falls, which it hath from the Mountains.

Of Biaſara.

Biaſara is a Kingdom in the lower *Ethiopia,* in the Gulf of St. *Thomas* (by our *Engliſh-*Seamen called the *Bight)* the people very Barbarous, their habits made of Mats, they adict themſelves to Witchcraft, and ſometimes ſacrificing their children to Devils.

Of Monomotapa.

THE *Monomotapa,* that is the Emperor King, or Soveraign of *Motapa;* and poſſeſſes an Empire ſo great, that it is 1000 Leagues Circuit, this Prince deports himſelf with gravity, and that there is no acceſs to his perſon, but with very great ſubmiſſions, he is alwayes adorned with Chains and precious ſtones like a woman;

The Inhabitants are all black, of mean Stature, and excellent good footmen, that they are ſaid to out-run horſes.

The Woods have great ſtore of Elephants as alſo other Beaſts; rich Paſtures well furniſhed with Cattle, plenty with Grains, Fruits, Fowl, and is well watered with Rivers, in which are abundance of Fiſh, the Air temporate, their ſummer is when we have our winter, and their winter is when we have our Summer.

Of

Of Africa.

Of the Abiffines.

THE Empire of the *Abiffines*, *Heylin* makes to be the Dominions of *Prester John*, and faith he is of such great force that he is able to raise upon a sudden occasion, a million of fighting men, his Government is absolutely Tyrannical; The people profess the Christian Religion, which was first made known to them by the Eunuch of Queen *Candace*, who was baptized by *Philip* the Evangelist, and more generally by the Preaching of *Saint Mathew* the Apostle; since which they have much swerved from the Purity of the true Religion, by their many Corrupt Opinions, they keep many of the Ceremonies of the old Law, they keep the seventh day Sabbath according to the fourth Commandment, they allow their Priests no yearly maintenance, neither will they suffer them to beg, they get their livelyhood by their own labour, they administer the Ordinance of the Lords Supper to Infants presently after they are Baptized, they Baptise themselves in Ponds and Lakes every Ephiphany day, supposing that to be the day that *John* Baptized Christ in *Jordan*.

Titles of the Emperor as Dr. *Heylin* doth inform us, are as followeth, *viz.*

Supream of his Kingdoms, and the beloved of God, the Pillar of Faith, sprung from the stock of *Judah*: The Son of *David*, the Son of *Solomon*, the Son of the Column of *Sion*, the Son of the seed of *Jacob*, the Son of *Mary*, the Son of *Naha*, after the flesh, the Son of St. *Peter* and St. *Paul*, after the Spirit. Emperour of the higher and lower *Ethiopia*, &c.

Of the Islands Belonging to AFRICA.

THose that are situate in the Occidental or *Atlantick* Ocean, may be Marshall'd into three Bodies of Islands, *viz.* the Islands of the *Azores*, the *Canaries* and the Islands of *Cape de Verde*.

Of the Islands of Azores.

THe *Azores* are nine in number, which are 1 St. *Michael*, 2. St. *Maries*, 3. *Tercera*, 4. St. *Gratiosa*, 5. St. *George*, 6. *Foial*, and 7. *Pico*, 8. *Flores*, and 9. *Corvo*. The Air of these Islands is generally good, they are well stored with Flesh, Fish and Fruits.

Of the Canary Islands.

THe *Canary Islands* are in number seven, The Names are as followeth, viz. 1. *Lancerotta*, 2. *Forteventura*, 3. *Canaria*, 4. *Teneriffa*, 5. *Palma* 6 *Ferro* and *Gomera*.

These Island abounds in excellent Wines, Honey, Wax, Sugar, Oade, Lawrel-trees, Dragon-trees, out of which they draw a Red Liquor, called by our Apothecaries *Sanguis Draconis*. Also Corn and Sugars, and all sorts of Fruits, stored well with Cattle, and is therefore the victualling place of the King of *Spains* Plate-Fleet.

To the Northward of these Islands lyeth the Island of *Madera*: belonging to the Crown of *Portugal*, The Air very wholsom, refresht with pleasant Fountains and Rivers, it yields also a pleasant Wine called *Madera* Wine.

The Islands of Cape de Verd.

THese Islands are in number ten, bearing the name of 1. St. *Mayo*, 2. St. *Anthonyes*, 3. St. *Vincent*, 4. St. *Luce*, 5. St. *Nicholas*, 6. Isle *de Sal*, 7. *Bona Vista*, 8. Isle *de Fuogo*, 9. *Brava*, 10. St. *Jago*; Some of these Islands yields good store of Salt, but the Air of them is not very wholsome.

Princes Island yields Fruit, Sugar, and Ginger;

and

and is the Revenue of the Prince of *Portugals* and therefore called *Princes Island*.

Annobon yields Sugars, Cottons, Cattle, and Excellent Fruits, and Oranges. Inhabited by *Portugalls*.

St. Helena lying in sixteen degrees south latitude belonging to the *East-India* Company, well furnished with good water, which alone is a great refreshment to the Ships that return from *India*; the Valleyes are very Fertile, the Air healthful, that sick persons are in a short time restored to their health; There are also in this Sea the Islands of Ascention, and *St. Mathewes*, affording Fowls and Wild Beasts, but not Inhabited.

Zocatora and *Babel Mandel* lies towards the Red Sea, where the passage is narrow and most convenient from the Coast of *Africa*, *Zocatora*, near *Cape Gauderfu* is under the Jurisdiction of an *Arabian* King, it is a good Road, and hath convenient Bays where Ships may ride secure among the very Rocks. It affords excellent Fishing, Cattle in great abundance, and is famous for the quantity and goodness of its Aloes.

Of Madagascar.

THe Island of *Madagascar* is the bigest of all the Islands that belong to *Africa*; The Air is temperate, the soyl produces several sorts of Grains and Trees, the Waters excellent, the Fruits delicious. The Mountains are full of wood, Pasturage and Plants, &c. The Natives very black, and of Rude Behaviour, partaking of the Customs and Manners of the *Africans*.

A General Description OF AMERICA.

CHAP. I.

AMERICA, fo called from one *Americus Vefpucius* (who following the Steps and Examples of *Columbus* and *Cabot*) Difcovered a part of this great Continent, which might as properly have been called *Columbana*, *Sebaſtiana*, or *Cabotia*, but moſt improperly the *Weſt Indies*.

It is bounded on the Eaſt with the *Atlantick Ocean*, on the Weſt with the *Pacifick Ocean*, on the South with *Terra Auſtralis Incognita*, from which it is feparated by the Streights of *Magellan*: The North bounds of it, not hitherto fo well known as that we can certainly affirm it to be an Iſland or Continent.

The Natives are Fair and Clear, little inclining unto Blackneſs, being generally Tall and well proportioned, their Eys little and black, ſtrong and healthful; for the moſt part Naked, unleſs a Cloth about their Waſte.

Their Language high and lofty in signification, for one Word serveth instead of two or three, the rest are supplied by the understanding of the hearer.

Their Houses are Mats or Bark of Trees set on Poles, in a fashion of our *English Barnes*, they lie on Reeds or Grass: As to the other Rights and Customs, I shall mention in the respective place.

For their Original, 'tis supposed to be of the *Jewish Race*, that is of the Stock of the *Ten Tribes*, and that for the Reasons following, 1. They were to go to a Land not planted or known. 2. Their Countenances and Children resemble the *Jews*. 3. They also agree in several Rites and Ceremonies, for they reckon by the Moon. 4. They Offer their first Fruits. 5. Many words they have of the *Hebrew Ideom* amongst them. 6. They have a kind of Feast of *Tabernacles*. 7. They are said to Build their *Altar* upon Twelve Stones. 8. Their Mourning is a Year. 10. Customs of Women, as to their Separation from their Husbands, after the manner of the Law of *Moses*.

It is commonly divided into Two Parts, viz, into South and North *America*, the several *Colonies* thereof take as follow.

CHAP. II.

Of South America *in Particular.*

Of Magellanick Land.

IT lies upon the Southernmost part of *America*, near the Streights of *Magellan*, whose Name it still bears: It is a very poor Countrey, much subject to Cold;

of AMERICA.

Cold; the Natives live in Caves: We have the Names of some places; as 1. *Desaguadore*, and 2. *Magellanick*; but having no perfect knowledge of them, I can say little thereto.

Of Paraguay.

PAraguay or rather *Plata*, so called by reason of a River of that Name that Waters it, the Countrey is very pleasant and delightful, for it abounds in Corn, Vineyards, Fruit Trees, and Cattle in abundance; places of most note are 1. St. *Jago D'estra*, 2. *Villa Rica*, and 3. St. *Anne*.

Of Chili.

CHili bears the Name of one of her Valleys, much subject to Cold, yet in some parts the Soil is so fertile and pleasant, that no part in all *America* more resembles *Europe*; it yields Ostriches, Copper, and the finest Gold in the World. St. *Jago*, *Imperiale*, *Baldavia*, and *Castro* are the principal places of *Chili*.

Of Peru.

PEru, though it gives Name to all the *South America*, yet it is but meanly furnished with Food, the chiefest thing being *Maize*, which is not very

very Plentiful: The Commodities are Gold, Cotten, and some Medicinal Drugs; it is divided into 1. *Quiro*, 2. *Truxillo*, 3. *Lima*, 4. *Cusco*, and 5. *Arica*, of which we have no other Relation than what the *Spaniards* reports.

Of Brasil.

THough it lies under the *Torrid Zone*, neverthelesi the Air is Temperate and Fertile: The Commodities besides Brasil, are Amber, Balsom, Tobacco, Train-Oyl, Cattle of divers sorts, Sweet-Meats, and Sugar in abundance: It is divided into several *Capitanies*, as 1. *Siara*, 2. *Saltan*, 3. *Para*, 4. *Paraiba*. 5. *Pernambuco*, 6. *St. Salvador*, 7. *Ilheos*. 8. *Porto Segaro*, 9. *Spirito Sancto*, 10. *St. Sebastian*, and, 11. *St. Vincent*.

Of Amazones.

AMazones or *Guiana*, hath its Name from a River so called; the Air is Healthful, the Soil is good in some places; for Tillage of Maniac, Cotten, Sugar, Tobacco, Gums, Wood, Stones of divers sorts, Parrots, and Monkeys: Places of most Note are 1. *Coropa*, and 2. *Villago D'or*, but very little known to us by reason that the *Spaniards* suffer none besides their own Nation to come into the Countrey, but kill all strangers they find.

Of

Of Terra Firma.

OR *Caſtill del Ore*, the latter given by the *Caſtilians*; the chief places are 1. *Panama*, 2. *Cartagena*, 3. St. *Fed Bagota*, 4. *Venezuola*, 5. *Surranam*, and 6. *Manoa*. The Air is very unhealthful, the Commodities divers, as Balſom, Roſin, Gums, Long Pepper, Dragons Blood, Stones of divers ſorts, and Gold: They have ſeveral ſorts of Beaſts, as the *Viuves* or *Rams*, &c. which you may find in *Helyns* Coſmography.

The chief Rivers in this *Southern America* are 1. *Oronoque*, which overfloweth once a Year as doth the *Nilus*, 2. *Amazones*, the greateſt and ſwifteſt in all *America*, 3. St. *Franciſco*, 4. *Paraguay* or *Plata*, 5. *Uraguay*.

Of the Iſlands of AMERICA.

Of the Caribbees.

BEtween South and North *America* lies ſeveral *Iſlands*, the firſt are the *Carribee* or *Cannibal Iſlands*, which are ſeveral ſmall *Iſlands*, which lie extended from the Coaſt of *Paria* to the *Iſle Porto Rico*, the

the chief are, 1. *Granada*, 2. St. *Vincent*, 3. *Dominica*, 4. *Barbados*, 5 *Antego*, 6. St. *Christophers*, 7. *Nevis*, 8. *Monserat*, &c. The Air good, considering how they lie; the chief Commodities being Sugar, Cotten, Ginger, and Tobacco, Inhabited by several Nations.

Of the Lucaie Islands.

SO called from *Lucaion* the Name of the biggest; of little Note, unless for a Fountain, which is said to Renew Youth again, and for their handsome Women, of which they are reported to have great store.

Of Porto Rico.

THe Air is very Temperate and Pleasant, the Soil indifferent Fertile; the Commodities Sugar, Ginger, and Cassia.

Of Barmudas.

IT is an *Isle* of a good Temperature, the Soil Fertile and Good, yielding Two Crops a Year, having excellent Fruits; the Commodties are Sugar, Oranges, Cochaneel, and Tobacco, and some Cotten also they have, but no great store. It is subject to the Crown of *England*.

Of

Of Jamaica.

THe Soil Rich and Fat, the Trees and Plants being always green and pleasant; the Air more temperate than any of the other *Isles*: The Commodities besides Sugar, Cotten, Indico, and Tobacco, are divers and plenty; they have Cattle, Fowl, Fish and Fruits of divers sorts. It is subject to the King of *England*.

Of Cuba.

THe Air is temperate and good, the Soil fertile, the Commodities are Ginger, Cassia, Mastich, Aloes, Cinamon, and Sugar, also Gold, but somewhat drossy.

Of Hispaniola.

THis is much like *Cuba*, saving that the Gold is more pure without Dross. It is subject to the King of *Spain*; somewhat Hot and Unhealthful, much subject to Thunder and Lightning, by reason of its situation so near the Equinoctial.

CHAP. III.

Of North America *in Particular.*

Of New Mexico.

IT is a Province little known to the *Europians*; the Inhabitants being divers in Language, Manners, and Customs. It is divided into *New Mexico, Arian, Quiviria,* and *Libola.*

Of Mexico *or* New Spain.

A Conntrey enriched with innumerable Mines of Gold and Silver: The Air temperate, the Soil fertile and good, The chief Towns are 1. *Mexico*, 2. *Guatamala*, 3. *Truxillo*, 4. *Acupulco*, 5. *Panaco*, The Commodities besides Gold and Silver, are Copper, Iron, also Wooll, Silk, Sugar, and divers Medicinal Drugs. They have also several other Commodities, which are too long to insert in this place. It is fully subject to the *Spaniard*.

Of

Of Florida.

IT is a place of very good Temperature, the Soil very Fertile, full of Fruit-Trees; the Towns well peopled; yet the Coast is very inconvenient for great Vessels, by reason of the Shallowness of the Water: Places of most Note are *St. Martha* and *Cofa*.

Of Carolina.

IT is a Countrey blest with an excellent Temperature of Air, the Soil Rich and Fertile, producing excellent Fruits, the Earth also apt to bring to Maturity Corn, all sorts of Garden Herbs and Roots: The Commodities are Wines, Oyls, Silk, Cotten, Indico, Ginger, and Tobacco; plenty of Fish, Fowl, and Cattle; the chief Town is *Charles Town*, Governed by one at the Appointment of the Proprietors.

Of Virginia.

THe Air of this place is sufficiently pleasant, the Soil exceeding Fertil; it produceth all sorts of Grain and Pulse, divers sorts of Garden Herbs and

Roots

Roots ; Silk Worms alfo which make good Silk ; the Commodities divers, but the chief is Tobacco. The place of moſt Note is *James Town*, Governed by one Deputed by the King of *England*.

Of Penſilvania.

IT is a place not yet well Planted, but may be in time, the Soil and Air being fit for the Nature of an *Engliſhman*: Granted by Patent from his late Majeſty King *Charles the Second*, unto *William Penn* Eſq; and his Heirs for ever, and therefore caled *Penſilvania*.

Of Mary-land.

HAving given you ſo full an account of *Virginia*, I need ſay little more, only that the general way of Commerce in both places being by interchanging one Commodity for another, and that which ſetteth a Price upon all other is Tobacco, there being ſuch abundance of this Imported into *England*, that the King hath 60000 *l. per Annum* for *Exciſe* and *Cuſtom*. The chief Town is *Baltamore*.

Of New Jersey.

FOr Temperature of Air and Fertility, there hath been enough faid already in *Virginia* and *Mary-land*, this place partaking of all the Properties and Advantages of them both.

Of New York.

A Colony fo called from his Royal Highnefs the Duke of *York* our prefent King. A Countrey found to produce the fame Birds, Beafts, Fifhes, and Fruits with *New England*, being Rich and Fertile. The chief Town *James Town*.

Of New England.

IT is a vaft Tract of Land, healthfully feated, the Soil exceeding Fertile, for it yields Wheat, Rye, Peafe, Beans, Barley, Oats, Indian Corn, Flax, Hemp, and all forts of *Englifh* Herbs. It hath plenty of Cattle of divers forts, Fifh, Fowl, and good Cyder. It excels with good Cellarage to preferve all, which is not common in *Virginia*; the chief

chief Commodities are Furs, Flax, Amber, Iron, Pitch, Tar. Masts and Timber to Build Ships. Their Metropolis is *Boston*, well seated, and adorned with fair and beautiful Houses, and well peopled.

Of New Scotland, New France, *and* Canada.

THree Places full of Stags, Bears, Martens, Hares, Foxes, and store of Conies, Fowl, and Fish; not over Fertile. The chief places are *Port Royal* in *New Scotland*, *Quebeck* in *New France*, and *Brest* in *Canada*, of no great Importance.

Of New Britain, New South Wales, *and* New North Wales.

THree Provinces much like the former (we having but little knowledge thereof) only the Soil is somewhat better. Places of most note are *Fort Charles* in *New Britain*, *Port Nelson* in *South Wales*, and *Ne Ultra* in *North Wales*. It was in some of these places that *Hudson* and others Wintered in their Voyages to the North West.

Of

Of the Island of California.

THe Air hereof is indifferently Temperate, being full of Herbage and Cattle, which be little less than them of *Europe*; supposed to have some Traffick with *China*, but not certainly known.

Of Newfound-Land.

IS an *Island* famous for its Bays, Harbours, and the great store of Fish caught there; and therefore much frequented by *French*, *Dutch*, and *Biscaners*; some part thereof Granted by Patent to Sir *George Calvert*, and still possessed by his Son and Heir the Lord *Baltamore*.

Of Groenland.

GRoenland contains a vast Tract of Land, not yet fully discovered, though it hath been long known to the *Norwegians*, who have several Colonies

nies planted therein; it hath alfo been touched at by feveral of our *Englifh Men* in their Voyages to the Northweft: The Sea hath great ftore of Whales, alfo Sea Wolves, Dogs, and Calves, and White Bears, which are faid to live more by Water than Land, and Feed moft upon Fifh: they have alfo Wheat, Chefnuts, Apples, and good Grafs for Pafture: There are alfo Horfes. Stags, Wolves, Foxes, Dogs, and Martles. If I fhould go about to relate all the Stories (which are more ftrange than true) that are told of the ftrange things in this Countrey, I fhould need a far larger Volume than this to contain them.

Of Ifland.

ISland is an 150 Miles long, and little lefs than an 100 broad. Its Inhabitants are fubject to the King of *Denmark*; fo healthy are they, that they live to an hundred years of Age; neverthelefs very unlearned, and little are they skilled in the Liberal Sciences, following nothing fo much as the Feeding of their Flocks and Cattle.

In this *Ifland* are Two Mountains, the Name of one is *Hecla*, the other *Helga*, which vomit and fend forth Fire in abundance: the firft being fo fierce, that there is no approaching it by fix Miles, and therefore the place is much indamaged thereby, that it is a great lofs to the Inhabitants.

Of Hudsons Bay.

BEtween *Groenland* and the Coast of *Nova Francia*, lieth a great Sea called *Hudson's Streights*, which after some Leagues Passage openeth into a fine Bay, dilating it self both toward the North, South, and West, giving great hopes of a passage that way to the *East Indies*. First discovered by the Two *Cabots*, Father and Son, upon the account of *Henry the Seventh* of *England*. Afterwards by *Hudson, Forbisher, Weymouth, Button, Baffin, Smith, James, &c.* and of latter times we have had a Traffick thither, which is managed by Merchants of *London*, called by the Name of *Merchant Adventurers to Hudsons Bay*: the Coast of *New Britain* runs along by the side of some part of this Bay. and though it has been much sought into, yet it cannot be found out yet whether there be a passage this way, or whether it is no more but an Inlet of the Sea; there being several places called by the Name of the first Discoverers, as the place where *Hudson* Wintered *James Bay*, *Buttens Bay*, *Forbishers Streights*, *Freum Davis*, *Baffins Bay*, &c.

Thus have I given you as full a Relation of *America* as the bulk of my Book will permit (being nothing (according to the best of my knowledge (but what

what is the true state thereof at this time, hoping that if there be any mistakes you will not impute it to my carelesness, but to the Mif-information of them that have writ of any of these New Discoveries. I shall end therefore with that Advice of *Helyn*:

But whither goes my Bark? Return: for we
Have flic'd the Capering Brine enough: See, see
The South Wind 'gins to gather Clouds apace;
'Tis no safe tarrying in so fierce a place.
Whilst thou hast time, retire, thou wearied Bark
Into safe Harbour; when the Cloud which dark
The Worlds bright Eye shall be dispell'd away,
And shining Phœbus make a lightsome Day.
Tritons shrill Trump shall thee recall again,
From the safe Harbour to the foaming Main,
And we with all our Powers will boldly try
What of this Unknown World we can descry.

F I N I S.

All sorts of Mathematical Books and Instruments both for Sea and Land; Maps both great and small of all Countreys, Maritine Charts, and Sea Plats, are sold by *John Seller* at his Shop on the West-side of the *Royal Exchange* in *London*, and at the *Hermitage* in *Wapping*.

Scales of Miles and Leagues, of Diverse Nations, shewing what Proportion they bear to a degree of the Earth.

The Length of one degree.
10 20 30 40 50 60

Roman, Italian, Turkish, and English miles, 60 to one degree
10 20 30 40 50 60

Arabian, English, and French Leagues, 20 to one degree or hours goeing
5 10 15 20

Spanish Leagues 17½ to one degree
5 10 15 17½

Common German miles 15 to one degree
5 10 15

Swedes and Danish miles 10 to one degree
5 10

Hungarian miles 14 to one degree
2 4 6 8 10 12 14

Scotch miles 40 to one degree
10 20 30 40

Low-dutch or Hollands miles 19 to one degree
5   10   15   19

Russian miles 80 to one degree
10 20 30 40 50 60 70 80

Persian miles 8 to one degree
1 2 3 4 5 6 7 8

East Indian miles 100 to one degree
10 20 30 40 50 60 70 80 90 100

Kingdom of Cambaja, & Gazuratt, 30 Cosa's to one degree
5 10 15 20 25 30

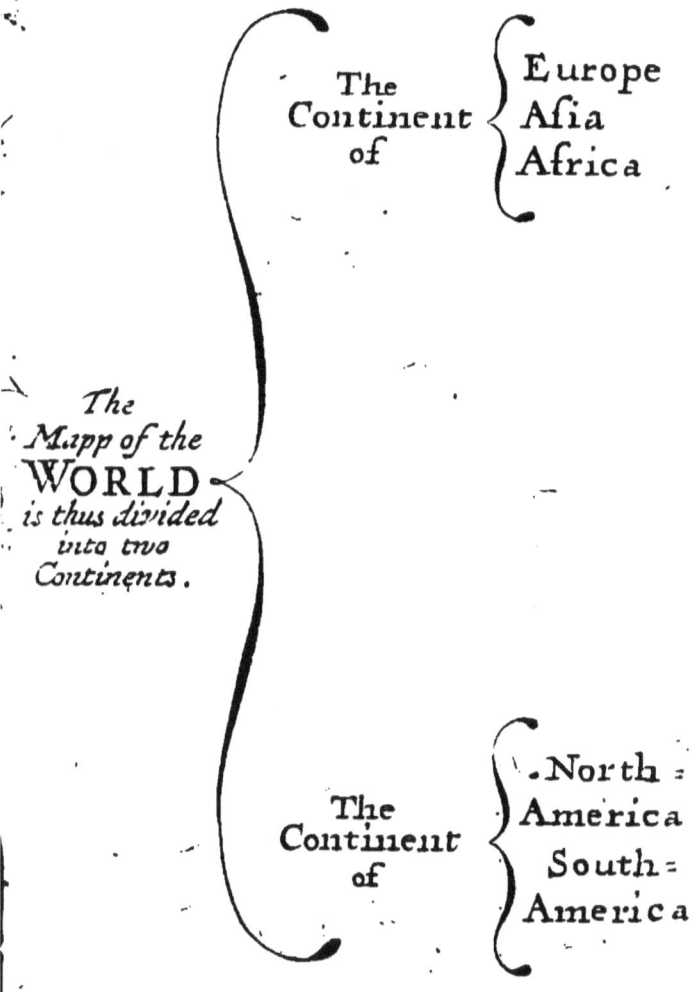

The Mapp of the WORLD is thus divided into two Continents.
- The Continent of
  - Europe
  - Asia
  - Africa
- The Continent of
  - North America
  - South America

A NEW MAPP

OF THE WORLD.

North

EUROPE is thus Divided

- England — London
- Scotland — Edinburg
- Ireland — Dublin
- Spaine
  - Madrid
  - Mallaga
  - Bilboa
  - Gibralter
- Portugal — Lisbon
- France
  - Paris : Brest
  - Marsselles
- Italy
  - Rome
  - Venice
  - Genoa
- Germany
  - Vienne
  - Prague
  - Hamburg
- XVII Provinces — Amsterdam
- Norway
  - Bergen
  - Drontem
- Sweden
  - Stockholm
  - Riga. Abo
- Denmark
  - Rypen
  - Copenhaven
- Poland
  - Danzick
  - Cracow
- Lithuania — Wilna
- Moscovia or Russia
  - Moscou
  - Archangel
  - Wologda
  - Cazan. Kola
- Lesser Tartary — Caffa
- Turky in Europe
  - Constantinople
  - Buda
  - Ragusa

The chiefe Rivers are
- Danube. Rhin
- Loyre

The Kingdom of ENGLAND is divided into two parts.

- ENGLAND is divided into forty Countyes:
  - Buckingham shire
  - Bedford S: Berk S:
  - Cambridg shire
  - Cheshire. Cornwal
  - Cumberland
  - Darby S: Devon S:
  - Dorset S: Durham
  - Essex. Gloucester S:
  - Hereford shire
  - Hant S: Hertford S:
  - Huntington S: Kent
  - Leicestershire
  - Lancashire
  - Lincolnshire
  - Monmouthshire
  - Midlesex. Norfolk
  - Northampton sh:
  - Northumberland
  - Nottingham shire
  - Oxford S: Rutland
  - Shropshire. Sussex
  - Somerset S: Suffolk
  - Stafford S: Surrey
  - Warwick S: Wilt S:
  - Westmorland
  - Worcester S: York S:

- WALES is divided into twelve Countyes:
  - Anglesey
  - Brecknockshire
  - Cardiganshire
  - Carmarthenshire
  - Carnarvanshire
  - Denby S: Flintshire
  - Glamorganshire
  - Merionethshire
  - Montgomery S:
  - Pembrook S: Radnor.

SCOTLAND is thus Divided.

In the North are these Divisions:
- Cathanes
- Strath-navern
- Southerland
- Assynt-ross
- Lochquaber
- Murray
- Anie beyne
- Buchan : Marr
- Badenorth
- Lochabvr
- Anthol
- Gowre : Mernes
- Angus
- Perth
- Broad Albain
- Argile : Lenox
- Strath
- Menteith
- Eise : Lorne
- Cantyr

In the South are these Divisions:
- Sterling
- Reinfrew
- Cuningham
- Lothia
- Clwydesdale
- Kyle
- Carrick
- Twedale
- March
- Tivedale
- Lidesdale
- Eskeda
- Annadale
- Nythdale
- Galloway

With many Islands

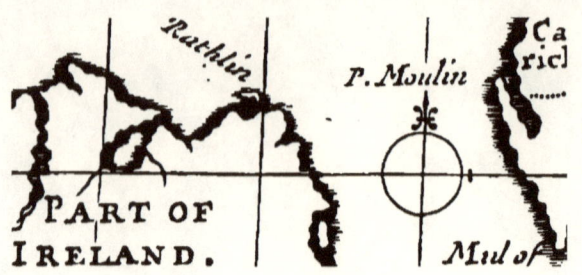

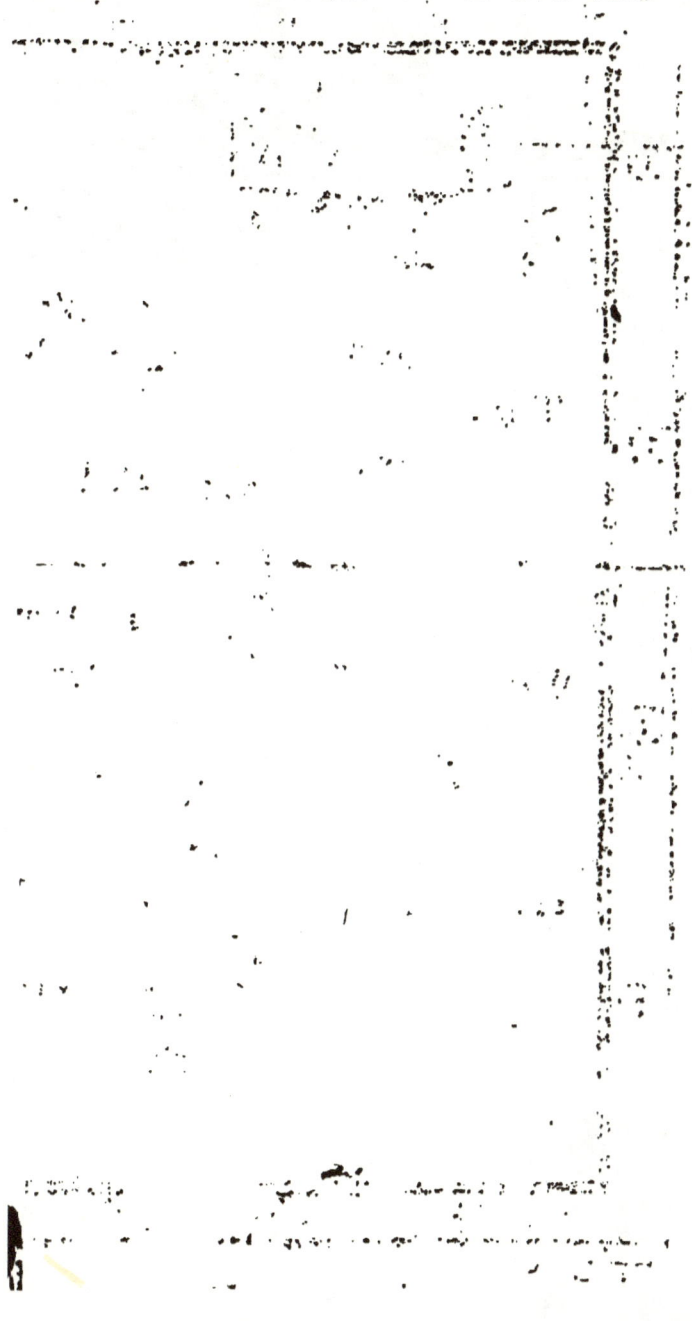

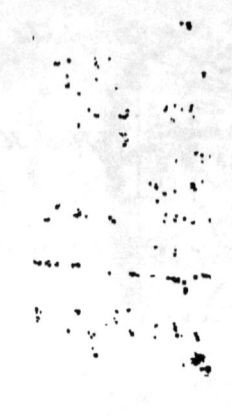

SPAIN is thus divided
- Biscaie
  - Bilboa
  - S. Sebastian
- Asturies
  - Oviedo
  - Santillana
- Galice
  - Coruña
  - Compostella
  - Tuy
- Leon
  - Leon
  - Salamanca
- Portugal
  - Braga
  - Port
  - Coimbra
  - Lisbon
  - Evora
  - Baja
- Algarve Faro
- Castile
  - old
    - Valladolid
    - Villa Franca
  - new
    - Placentia
    - Madrid
    - Toledo
    - Cuenca
    - Badajos
- Andalusia
  - Cordoue
  - Cadis
  - Seville
  - St Lucar
  - Gibralter
- Granada
  - Granada
  - Malaga
- Murcia
  - Murcia
  - Cartagene
- Valencia
  - Valencia
  - Alicante
- Arragon
  - Caragoça
  - Calatajud
  - Albarazin
- Catalonie
  - Lerida
  - Barcelona
  - Girone
  - Tarragona
  - Tortose
- Roussillon Perpignan
- Navarre
  - Pamplona
  - Estella

The Chiefe Rivers are ye
- Douero
- Tage
- Guadiana
- Guadalquivir
- Xucar
- Ebro

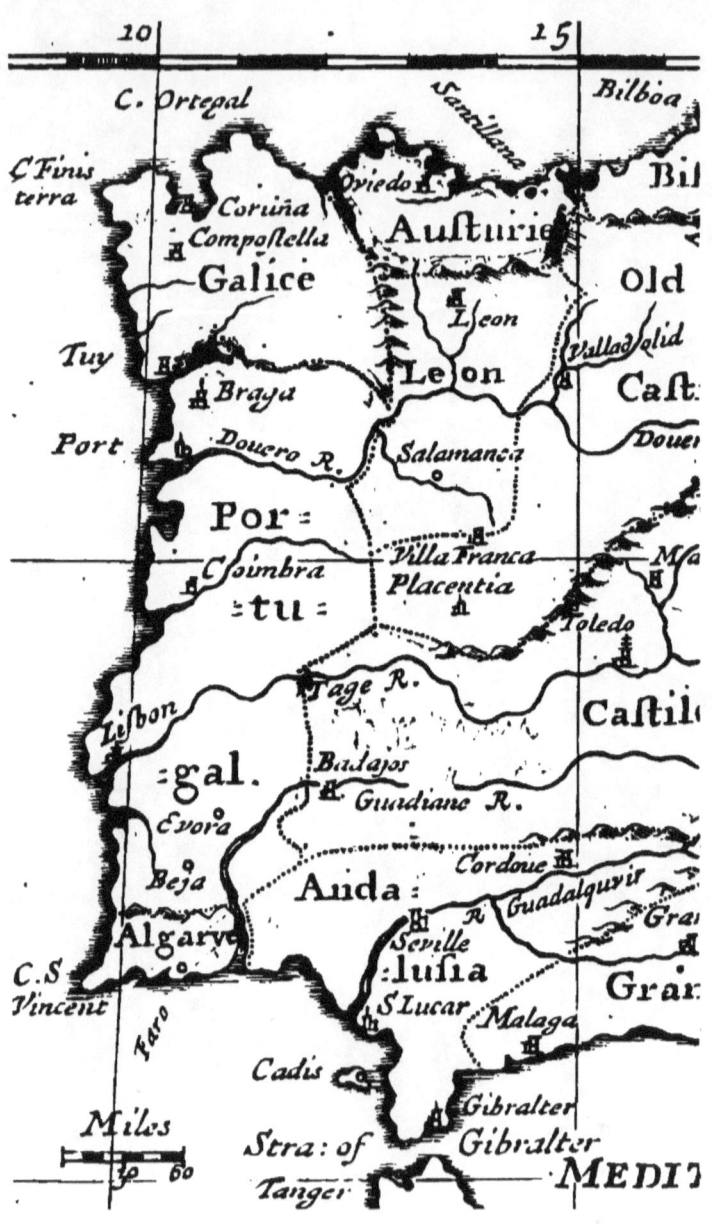

PORTUGAL is thus Divided
- Entre DouroMinho
  - Valence
  - Braga
  - Porta
- Tralos Montes
  - Braganca
  - Miranda
  - Moncorvo
  - Almeida
- Beyra
  - Lamego
  - Viseu
  - Mira
  - Coimbra
  - Sabugal
  - Castelbranco
- Estremadura
  - Lisbone
  - C. de Roca
  - N. Lisboa
  - Leiria
  - Santare
  - Tomar
  - Punhete
  - Almerin
  - Palmela
  - Serual
  - C. St Ioan
  - Alcacer de Sal
- Alentejo
  - Portalegre
  - Evora
  - Beja
  - S. Iago d'Cacem
  - Mertola
- Algarvia
  - Silues
  - Lagos
  - Faro
  - C. d' St Vincent

The chiefe Rivers are ye
- Douer
- Tage
- Gaudiana

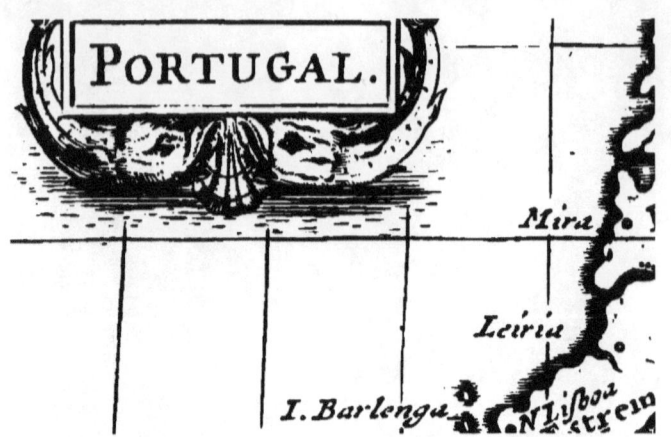

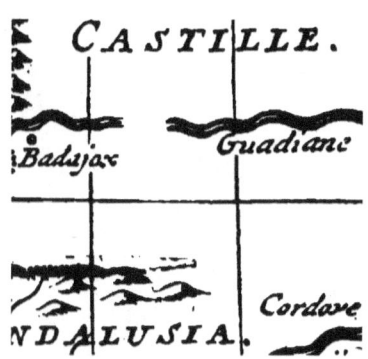

FRANCE is devided into these Provinces.

- Picardi
  - Calais
  - Amiens
- Normandie
  - Diepe
  - Haver de Grace
  - Rouen
  - Caën
- Bretagne
  - St Mallo
  - Rhennes
  - Nantes
  - Brest
- Orleanois
  - Orleans
  - Tours. Bourges
  - Angers
  - Poictours
  - la Rochelle
- Guienne
  - Bourdeaux
  - Rodes
- Gascogne
  - Aux
  - Bayone
  - Pau
- Languadoc
  - Thoulouse
  - Monpilier
  - Narbone
- Provence
  - Marselles
- Dauphine
  - Vienne
  - Grenoble
  - Embrun
- Lyonnois
  - Lyon
  - Clermont
  - Moulins
- Burgundy
  - Dyon
  - Auton
- Chāpagne
  - Tryers
  - Rhens
- Ille of France — Paris

The Chiefe Rivers are ye
- Seine. Loire. Dordogne. Adour.
- Garonne. Rhosne. Lot

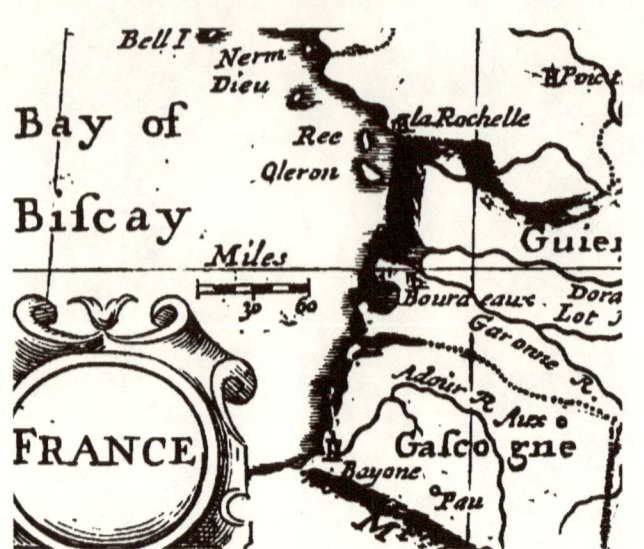

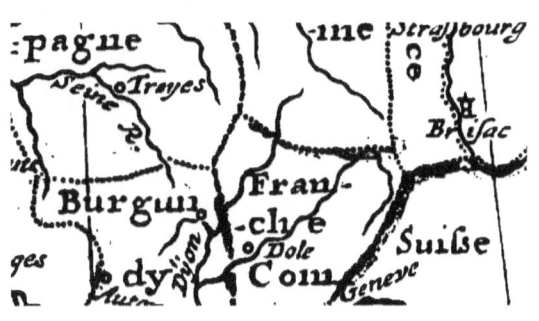

ITALY
is divided
into
these parts.
- K:m of Naples
  - Naples
  - Policastro
  - Regio
  - Gaeta
  - Tarante
  - Otranto
  - Brindisi
  - Potignano
  - Isola
  - Monfredonia
  - Vestica. Aquila
  - Pescara
- Estate of the Church
  - Rome
  - Spoleto
  - Fermo
  - Ancone
  - Urbino
  - Ravenne
  - Bologne
  - Ferrara
- Tuscane
  - Florenza
  - Ligorn
  - Pisa
  - Siena
- Luca ——— Luca
- Genoa ——— Genoa
- Parma ——— Parma
- Modena ——— Modena
- Montova ——— Montova
- Venice
  - Venice
  - Padoua
- Trente ——— Trente
- Milan ——— Milan
- Piemont
  - Turin
  - Nice

The Chiefe Rivers are ye
- Tiber
- Po

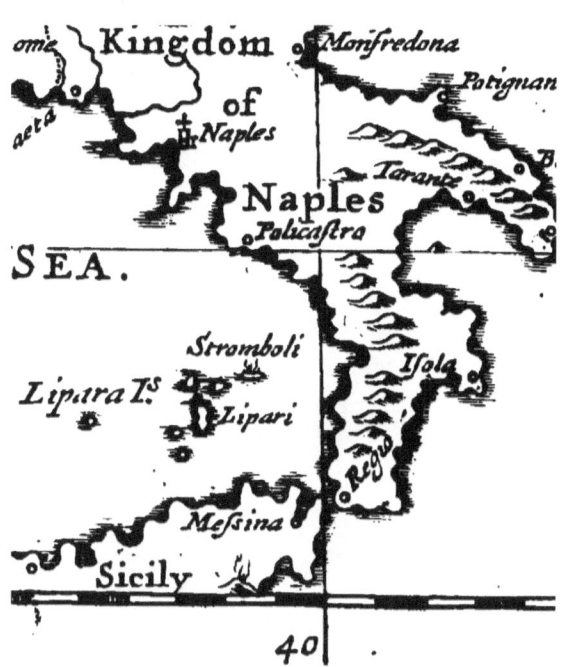

GERMANY is thus divided.

Meclembourg	Meklembourg, Wismar
Pomerain	Stralsund, Stetin, Coleburg, Rugen Isle
Brandenburg	Berlin, Kustrin, Francfort
Upper Saxony	Wittemberg, Erfort, Leypsick, Minden
Lower Saxony	Brunswick, Lubeck, Bremen, Hamburg
Westphalia	Emden, Munster, Paderborn
Hesse	Cassel, Marpurg
Ecclesiatick Electorats	Mauince, Cleves, Cologne, Treves
Palatinate of Rhine	Wormes, Heidlberg, Spire
Franconia	Francfort, Nuremberg
Loraine	Metz, Toul, Nanci
Alsace	Strasburg, Brisac
Franche Comte	Besancon, Dole
Savoy	Chambray
Suisses	Geneve, Berne, Basel
Sovabie	Augsburg, Ulm, Constance, Hailborn
Tirol	Inspruck, Landeck
Bavaria	Ratisbon, Passau, Salzbourg, Munich
Bohemia	Prague, Satz, Couigingracz
Lusace	Bautzen
Silesia	Breslaw, Oppelen, Teschou
Moravia	Obmutz, Bren
Austria	Vienna, Lincz, Stein
Stiria	Gretz
Carinthia	S. Veit
Carniola	Laubach

GERMANY.

POLAND.

Oppeln
Oder
Teschen
Cracow

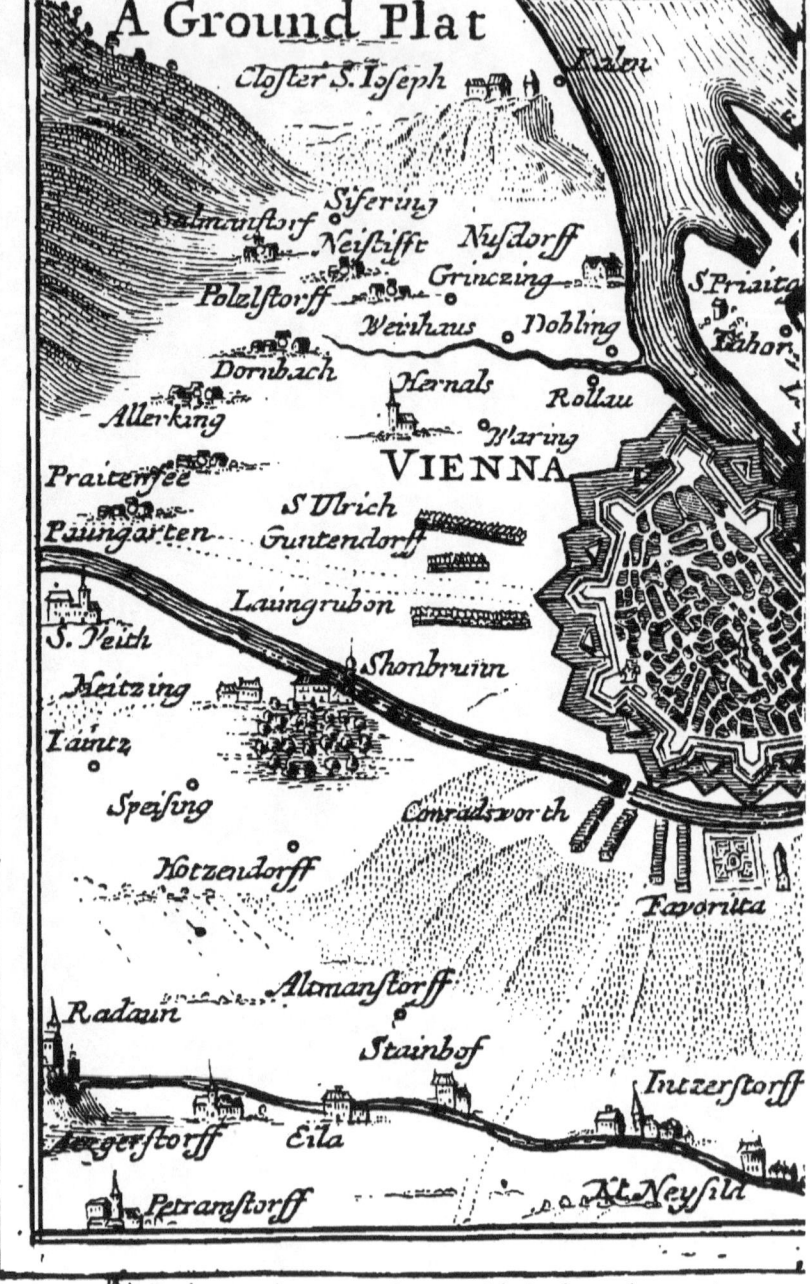

The City VIENNA and the Adjacent Country

HUNGARIA
The Greatest part is Conquered by the Turkes.

Upper Hungaria
- Presbourg
- Tranchin
- Nitria. Bars
- Cassovia. Rab.
- Epires. Papa
- Unghuar
- Rena. Sopron
- Tokoy. Vespron
- Namyn. Bator
- Ujogh. Zarmar
- Benhalora Comora

Lower Hungaria
- Newhausel
- Strigonie
- Alba Regalis or Stul Wessenbourg
- Offen. or Buda
- Pest. Vazzon
- Kanise. Lippa
- Agria. Schag
- Giula. Zeged
- Novigrod
- Zoluock
- Colocza
- 5 Eclesie. Zygeth
- Bathmonster
- Thurmy
- Waradin
- Czongrad
- Chonard
- Temesuar
- Breczkerk
- Mesasomlo

Sclavonia
- Posega
- Marsa. Valpon
- Waraßdin
- Szerem

THE XVII PROVINCES is Divided into
- **Dukedoms**
  - Limbourg — Mastrich, Limbourg, Masewick, Leive
  - Luxembourg — Luxembourg, Rochefort
  - Gueldre — Gueldre, Arnheim, Harderwick
  - Brabant — Charleroy, Boisleduck, Louvain, Breda, Brussels, Tillemont
- Marquisate of ye Holy Empire — Antwerp
- **Earldoms**
  - Flanders — Hulst, Gaunt, Bruges, Ipres, Mont Castlel, Dunkirque, Ostend, Lille, Tournay
  - Artoys — Arras, St Omer
  - Hainault — Mons
  - Namur — Namur
  - Zutphen — Zutphen
  - Holland — Amsterdam, Rotterdam, Leyden, Hage, Delf, Dort, Harlem, Brill, Edam, Horn
  - Zeland — Middleburg
- **Baronies**
  - Frisia — Lewarden
  - Utrech — Utrech
  - Overisle — Couwarden, Oldenzel
  - Malines — Malines
  - Groningen — Groningen

THE XVII PROVINCES

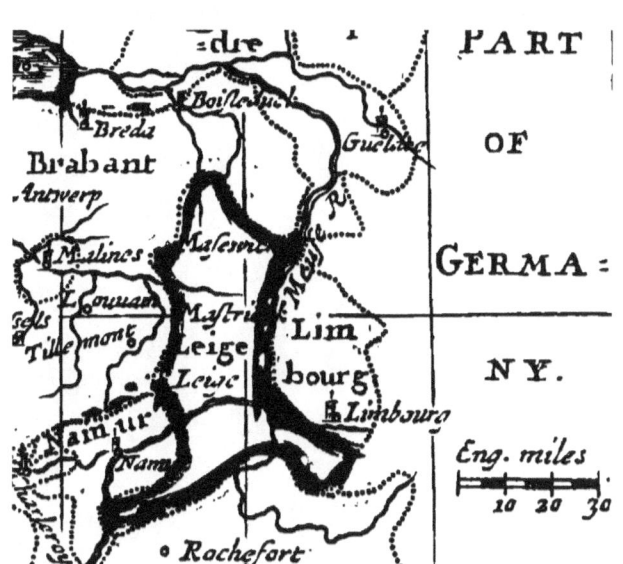

SWEDELAND and NORWAY is thus divided.
- Swead and Lapland
  - Stockholm, Ulm
  - Upsale, Torne
  - Nicopin, Kimi
  - Gevalie, Lula
  - Hundinkvald
  - Indal, Pitha
  - Hernsand
- Gothland
  - Gottenbourg
  - Elsenborg
  - Calmar
  - Norcopin
  - Carolstat
  - Ahuys, Bahus
- Livonia
  - Riga, Pernau
  - Revil, Derpe
  - Nerva
- Ingria
  - Noteborg
- Finland
  - Wiborg, Abo
  - Raseborg
  - Nitslot
  - Bienburg
  - Vasa
  - Oulo
  - Cayaneborg
- Norway
  - Fredrickstat
  - Obslo
  - Christiana
  - Bergen
  - Stafanger
  - The Nasse
  - Drontem
  - Salten
  - Hereles
- Norway Lapland
  - North cape
  - Wardhuyse
  - Roverda

.R.K. {
- Iutland {
  - The Scaw
  - Wensissel
  - Seeby
  - Alborch
  - Wiborg
  - Lemwick
  - Arhusen
  - Ebelted
  - Horsens
  - Yard
  - Henneborch
  - Rypen
- Sleswick {
  - Sleswick
  - Tonderen
  - Apenrad
  - Flensborg
  - Rensborg
- Holstein {
  - Neldorp
  - Kiel
  - Niemunster
  - Gluckstat
  - Oldeslo
  - Nieftat
  - Oldenborg
- Zeland I. {
  - Copenhage
  - Roskil
  - Elseneur
  - Nesned
  - Prestoe
  - Holbeck
- Fionie I. {
  - Odensee
  - Forburg

With severall small Islands

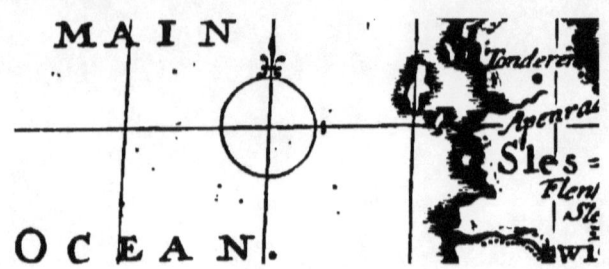

POLAND is thus Divided
- Great Poland { Gnesne, Posna, Kalisch, Rava
- Cujavia ------ Vladislau
- Lesser Poland { Cracow, Lublin, Sandomiri, Czeslacou
- Prussia Roy: { Danzick, Torne, Culm
- Prussia Du: { Koningsberg, Elbing, Maremburg
- Mazovia --- { Ploczke, Warsovia
- Polachia ---- { Tikasin, Bielski
- Black Russia - { Zamoski, Teroslau, Lemburg, Belz.
- Podolia { Kameniec, Braclau, Bar.
- Volhinia, Ukrain & Cosaques { Volodzimirez, Lusuc, Kiou, Krzemieniec, Zitomirs, Czernobel, Bialacerkiew, Czircassi, Kudac
- Lithuania { Braslau, Wilna, Kouno, Minski, Grodno, Novogrodek, Lakouvick, Orsa, Vitopski, Sklou, Rzeczica Mazi
- Curland --- { Vindau, Liba, Mitau
- Samogatie -- { Memel, Rosinie

LITHUANIA is Divided into these Palatinates & Duches.

- Poloczk — Poloczk. Drina.
- Witepsk — Witepsk, Wieliss, Surass
- Braslaw — Braslaw, Miudzia, Drysiwath
- Vilna — Wilna, Wilkomirz
- Troki — Troki, Lida, Kouno, Grodne
- Novogrodeck — Novogrodeck, Wolkowiska, Slonim, Ostrow
- Minski — Minski, Lessnica, Boryssow, Brodziec
- Mscislaw — Mscilaw, Mohilow, Bychow: Orssa, Balymissie
- Lands of Rohaczow & Rzeczyca — Rohaczow, Rzeczyca, Mazy, Dobossna
- Duche & Palat: of Smolensko — Smolensko
- Duche & Palat: of Novogrodeck Swierski — Novogrodeck Swierski: Starodub
- Duche of Czernihow — Czernihow, Sluczk.

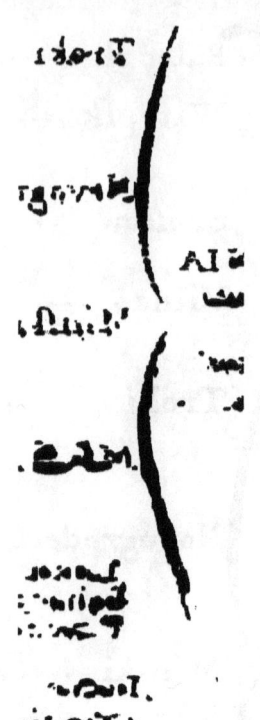

MOSCOVIA
or
RUSSIA
is divided
into.

- Kingdom's
  - Siberie
  - Cazan
  - Astracan
- Principalities
  - Pleskow
  - Bielskow
- Dukedom's
  - Novogorod Weliki
  - Nova Zemla
  - Obdora
  - Reschaw
  - Moscow
  - Twer
  - Belejezoro
  - Worotin
  - Ieroslaw
  - Wologda
  - Rosthow
  - Susdale
  - Wolodimer
  - T. de Mordwa
  - Rezan
  - Nisi Novogorod
  - Condora
  - Iuhorki
  - Permski
  - Waithka
  - Bulgar
  - Circasses Tartar
- Provinces
  - Dwina
  - Kargapol
  - Ustingha
  - Petzora
  - Okrain
  - Pole
- Republicks
  - Lapland
  - Samojedes
  - Tingoisis Manamo
  - Czeremissi Nagor:
    noi

1

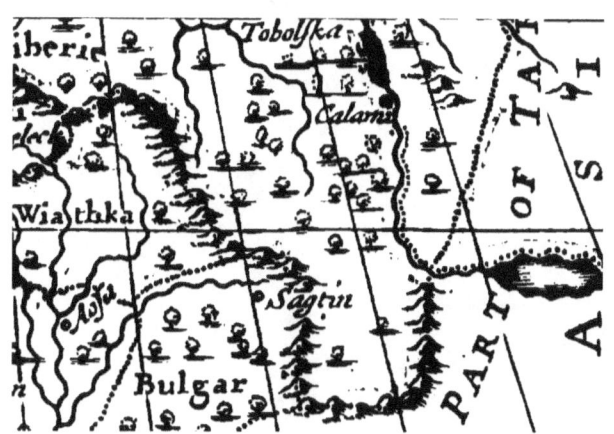

LESSER TARTARIA

In Lesser Tartaria are these Townes:
- Beyergenof
- Maniez
- Azac or Azow
- Pisan
- Paparoma
- Cambra
- Precop
- Bacuʃsarai
- Luʃtloua
- Mancup
- Baluclava
- Caffa
- Truʃta
- Carubas
- Kers
- Arbotka

On the Coast of PONTUS EUXINUS:
- Oczakou
- Buzlogrod
- Killia
- Constantinople
- Scutari
- Heraclia
- Sinopoli
- Simiʃo
- Trebiʃond
- Varth
- Fazo
- Savatopoli
- Eʃchiʃumuni
- Sophia
- Teman
- Temrok
- Cozala
- Bacmachi
- Baletecoi

TURKY in EUROPE divided.
- Upper Hungaria { Presbourg / Raab : Toky
- Lower Hungaria { Newhausel : Gran / Buda or Offen / Colocza. 5 Eclesiæ
- Sclavonia ...... Posega
- Transilvania ... { Clausenburg / Hermanstat
- Moldavia ...... { Soczowa / Iazi : Gallatz
- Walachia ...... { Ermstat / Torgowis
- Bessarabie ..... Bialigorod : Tekin
- Bulgaria .... { Uscopia : Sophia / Nicopoli : Varne / Tomi : Dora
- Servia .... Belgrade : Zizza
- Croatia .... Withitz
- Dalmatia ...... { Spalatra : Zoar / Raguse :
- Bosnia .... { Narenca : Luicza / Catoro : Risin
- Albania .... { Scutari / Durazo : Valone
- Romania ..... { Constantinople / Andrinopoli / Gallipoli / Asperosa
- Macedonia .... Salonichi : Contessa
- Thessalia ..... Larissa
- Epiros .... Perga
- Achaia .... { Lepanto : Negrepont / Setines : Stieus
- Morea .... { Corinte : Argos / Arcadia : Modon / Mississtra

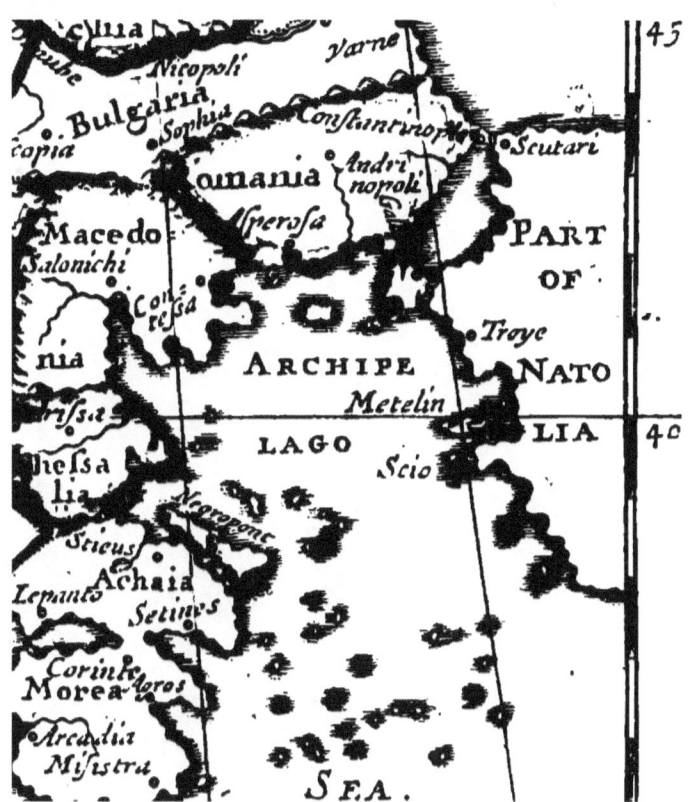

	Transilvania	Clausenburg Newmark Scespurg Meduuish Hermanstat Fogares. Egedin Deuua. Torda Huniad Hatzag
The North part of TURKY in Europe	Moldavia	Soczowa Sereth. Smatin Moldavia Tragorod Aczud. Iazy Rebnick Barlach Falxim Vasthuy Bradi. Tirasno Margosest Iapuczna Skoka. Galacz Srzepanovicze
	Walachia	Torgowis Barskow Pitesk. Arcun Rebnik Aluth. Zalatina Zula. Ris Zers. Zorlo Domboxisa Ermistat Ialonicz Brailonum
	Bess.arabie	Tekin. Orihou Bialigrod Moncastro Kilia. Smil Taristo.

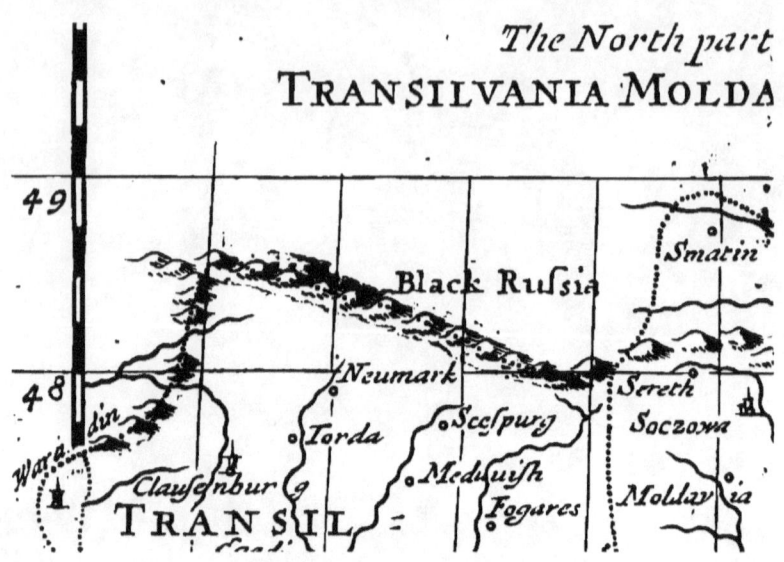

Turky in Europe
WALACHIA BESSARABIE.

The South part of TURKY in Europe
- Bulgaria: Uscopia, Ischa, Sophia, Nicopoli, Tomi, Varne, Dora
- Servia: Belgrade, Obrach, Semandria, Nizza, Noviba
- Croatia: Withitz
- Dalmatia: Zegen, Zara, Spalatra, Raguse
- Bosnia: Bagnaluc, Narenca, Iaicza, Risin, Catoro
- Albania: Scutari, Durazo, Valone
- Romania: Constantinople, Audrinopoli, Phillipopoli, Asperosa, Gallipoli, Chiorlich

GREECE
- Macedonia: Salonichi, Contessa, Pella, Acomania
- Thessalia: Larissa, Trica, Arimo
- Epiros: Butinito, Prevesa, Perga
- Achaia: Lepanto, Setines, Steius, Negrepont
- Morea: Corinte, Arcadia, Agros, Patra, Olimpe, Modon, Misistra

With many small ...... Islands

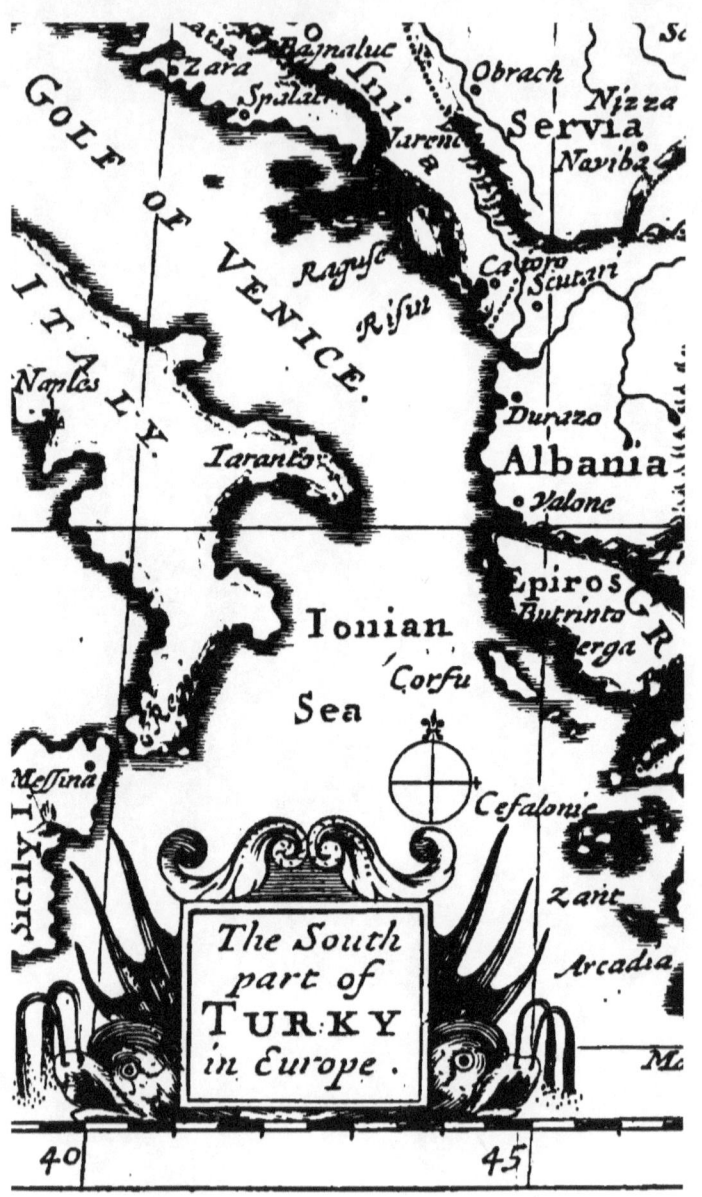

ASIA is thus divided.

- Turky in Asia
  - Ierusalem
  - Smirna
  - Bagdat
- Georgia
  - Cotatis
- Arabia
  - Mecca
  - Catif
  - Calajate
  - Fartach
  - Aden
- Persia
  - Ispahan
  - Taurus
  - Ormus
- Empire of Mogol
  - Lahor: Delly
  - Agra: Diu
  - Ougley
  - Bengala
  - Survat
- India *this side* Ganges
  - Bisnagar
  - Goa
  - Fort St George
- India *beyond* Ganges
  - Pegu: Sian
  - Camboja
  - Malacca
  - Tunquin
- China
  - Peking
  - Nanking
  - Canton
- Tartaria
  - Samarkand
  - Belch
  - Kasghar
  - Thibet
  - Tangut: Xamo
  - Chacan Kalmach

The Chiefe Rivers are the — — — —
- Ganges
- Ind. flu
- Eufrates

AFRICA is thus Divided.

- **Barbary**
  - Tanger
  - Morocco, Fez
  - Alger, Tunis
  - Tripoli
  - Barca

- **Biledul-gerid**
  - Tesset
  - Darha
  - Segelmesse
  - Tegoram
  - Tolacha
  - Guargala
  - Tenzara
  - Gaoga

- **Desart of Sarra**
  - Zanziga
  - Targa
  - Lempta
  - Bardoa
  - Borneo

- **Nigros**
  - Gualata
  - Genehoa
  - Tombut
  - Gambia
  - Cantori
  - Mandinga
  - Agades
  - Gago
  - Cano
  - Guber
  - Cassena
  - Zegzeg
  - Zanfara
  - Gangara

- **Guinea**

- **Congo**
  - S. G. d'Mina
  - Ardd
  - Benin
  - S. Salvador
  - Dongo
  - Loango

- **Biafara**
  - Biafara
  - Medra
  - Corisco

- **Monomotapa & Caffares**
  - Bagamedro
  - Monomotapa
  - Butua Zofala

- **Abissines**
  - Cucumo
  - Sova
  - Vangue
  - Angote
  - Ambiam
  - Amara
  - Damut

- **Zangubar**
  - Magadaxo
  - Adel
  - Malinda
  - Monbaza
  - Quiloa
  - Mozambique

- **Nubia** — Nubia
- **Egypt** — Cairo, Cosir

The chiefe Rivers are
- Nilus
- Niger

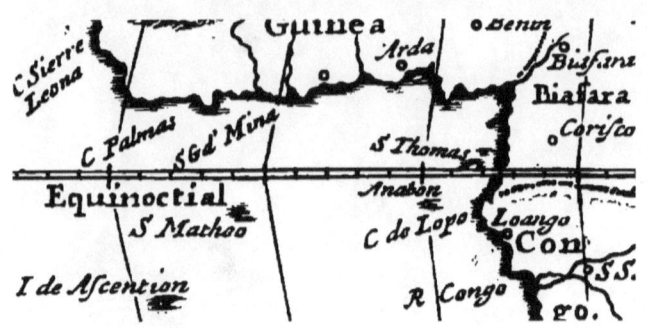

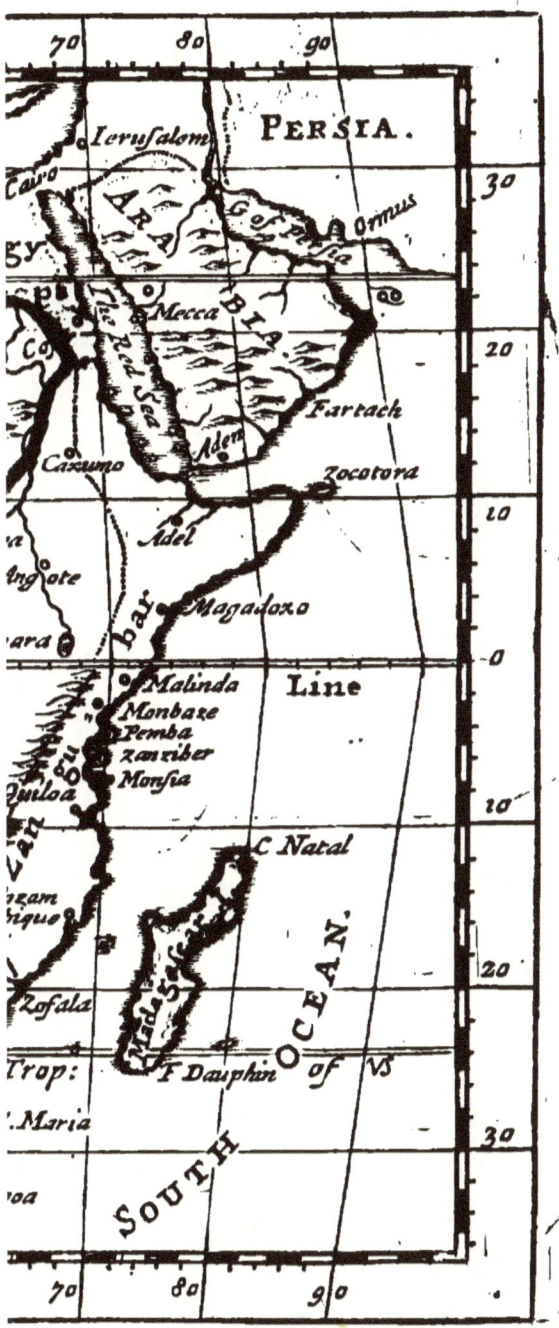

NORTH AMERICA *is thus divided*	New N. Wales	Ne Ultra
	New S. Wales	Port Nelson
	New Brittain	F. Charles
	Canada	
	New France	Quebeck
	New Scotland	P. Royall
	New England	Boston
	New York	New York
	New Jarsey	F. Elsenburg
	Maryland	Baltamore
	Pensilvania	
	Virginia	Iames Towne
	Carolina	Charles T.
	Florida	S. Martha / Cofa
	Mexico or New Spaine	Mexico / Guatamala / Truxillo / Acapulco / Panuco
	New Mexico	New Mexico
The chiefe Islands are		California / Hispaniola / Cuba: Long I. / Iamaica / Porto Rico / New Foundland / Barbados / Bermodas / Lucaie Islands / Caribes
The Great River		Canada

SOUTH AMERICA is divided into these parts.

- **Terra Firma**
  - Panama
  - Cartagena
  - St. Fe d. Bagota
  - Venezuela
  - Surranam
  - Manoa
- **Amazones**
  - Coropa
  - Village de lor
- **Brasil**
  - Para. Siara.
  - Saltpan
  - Paraiba
  - Pernambuco
  - St. Salvador
  - Ilheos
  - R.to Seguro
  - Sp.to Sancto
  - St. Sebastian
  - St. Vincent
- **Peru**
  - Quito
  - Truxillo
  - Lima. Cusco.
  - Arica
  - Potosi
- **Chili**
  - St. Iago
  - Imperial
  - Baldivia
  - Castro
- **Paraguay**
  - St. Iago d'estra
  - Villa Rica
  - St. Anna
- **Magellanick Land**
  - Desaguadero
  - Magellanick I.

The Chiefe Rivers are y.
- Oronoque
- Amazones
- St. Francisco
- Paraguay or Plate
- Uraguay.

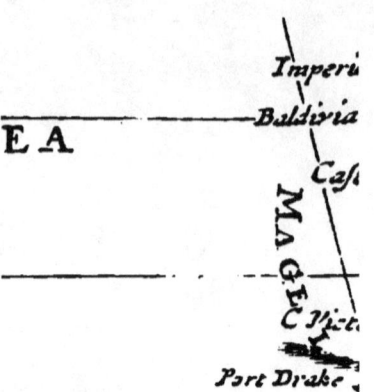

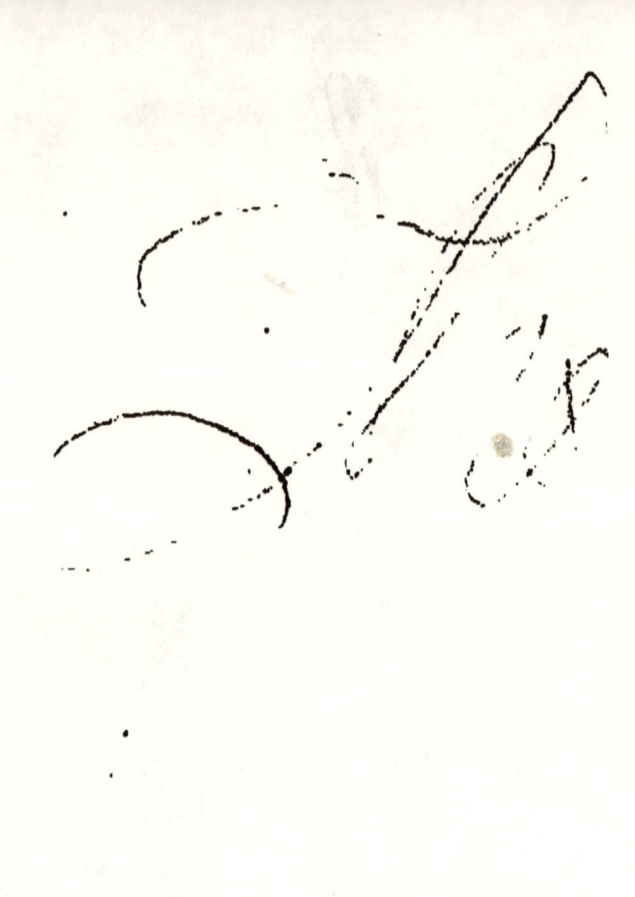

www.ingramcontent.com/pod-product-compliance
Lightning Source LLC
Chambersburg PA
CBHW032049230426
43672CB00009B/1530